DWELLING
IN A
NEW WORLD

Dwelling in a New World

Revealing a Life-Altering Technology

Robert Gold

iUniverse, Inc.
Bloomington

Dwelling in a New World
Revealing a Life-Altering Technology

iUniverse books may be ordered through booksellers or by contacting:

iUniverse
1663 Liberty Drive
Bloomington, IN 47403
www.iuniverse.com
1-800-Authors (1-800-288-4677)

ISBN: 978-1-4759-3075-7 (sc)
ISBN: 978-1-4759-3077-1 (e)
ISBN: 978-1-4759-3076-4 (dj)

Library of Congress Control Number: 2012910033

Printed in the United States of America

iUniverse rev. date: 07/03/2012

For the people of the world

CONTENTS

PREFACE

This work is nonfiction and atypical.

We hear many announcements of products that will be available shortly.

You likely have never received an announcement that your life will be drastically reoriented in only a number of months. It's also likely that you haven't heard about a transformation of consumerism and advertising such that they no longer occur as noise and actually address your immediate concerns.

Dwelling in a New World introduces you to a new technological concept. This technology obliterates linearity. Answering machines, texting, e-mails, TV programming, Google searches, computer programs, and other systems and schedulers as we know them will disappear.

In their place, a virtual world appears. Virtual companions support what is important to you, anticipate your needs, and acquire support from necessary resources.

Does this sound like fiction? Hardly. We have the technology, but it is currently designed to be something separate from us—as devices, programs, and tools we must use.

We are entering into a new technological era. In this world, we relate with ourselves and others in a clear, precise manner. As individuals, families, and organizations, we will have assurance of the future that has only been dreamed of and seen in futuristic adventures. Planning and supply and project management will exist in a completely new paradigm.

This book is intentionally short. It is vivid and expressive, intended to inspire your confidence in what life will be like in 2013 and beyond.

ACKNOWLEDGMENTS

There are my avid supporters, my incredible wife, Gayle, and our children. They were incredibly supportive and understanding. I appreciate my dear friend and her brilliant daughter, who assisted me by editing my fourth draft.

I have a special appreciation for those friends who argue and challenge me. They provided me a clarity that supported me in writing this book.

A very special thanks to a chairman of the Department of Sociology at a prestigious college and author of several scholarly books for his advice and comments on my second and third drafts.

INTRODUCTION

Whom this interests

Who is reading this?

If you are a person who desires to have full control over every aspect of your life, then this book will introduce you to a new technological environment that empowers your aspiration.

If you are someone who wants to simplify living to have comfort in life's most demanding moments and an awareness such that learning becomes a passion, then welcome to a new world in which to live.

If you are tired of cycles of excitement and disappointment and are resigned because it is hopeless to expect anything different, get ready for a life of new opportunities opening every day.

You might be a business executive constantly looking for a new, radical approach for

- ✓ automating and streamlining business practices (structures that eliminate processes);

- ✓ clarifying day-to-day interactions' influence on current and future resources and finances; and

- ✓ profoundly increasing productivity, effectiveness, and profitability, while lowering costs.

You might be a parent, sibling, friend, teacher, student, clergy member, scientist, professional, politician, or a retiree who wants fulfilling and satisfying lives for yourself and those closest to you.

As an inventive thinker, every moment that I live, everything that I do, think, desire, and experience ... every fiber of my being challenges the status quo. I am committed to altering our existence, to opening each of us to our hidden dreams and desires, and to tackling the demands and concerns of this planet. I firmly believe that a different realm for thinking is required to empower us.

The way that I challenge the status quo is straightforward. I find myself surrounded by complaints, fears, and longings for a different world. I discover the basic foundation of reasoning that developed many millennia ago. I identify the origin that shapes the confines—shapes how we see, act, and build things and ideas in the world. I remove it. A new world appears each time.

This is a book that opens up a new world. It announces an invention that rearranges our existing technology in such a way that it alters the world of technology itself. This alteration introduces a radical yet obvious reorientation that allows all of us to be aware of ourselves, others, and what is important in a completely different way.

In a world of dashed and unfulfilled expectations, one consistent complaint is undelivered promises. I understand this. If this complaint has resurfaced, I ask that, for the next few hours, you suspend disbelief.

From all my delights and disappointments, I have discovered that regardless of who you are or what your background is, you likely are interested in more effective communication, management of commitments (time), relationships, resources, activities, and money, while simultaneously experiencing a captivating and happy life.

This book is designed to be a very quick experience. Every few paragraphs, questions based on the current discussion will permit you to direct your attention to what is next.

Some of your dissatisfactions might instill strong views about the current state of our world, your family, communication, technology, accountability, or organizations you are involved with. You might find your current life hectic, full of distractions and interruptions. You might desire that your connection with people, plans, and what is important to you become simplified.

The author, book, and invention

Who wrote this? Robert Gold is a

- ✓ father and husband; and

- ✓ ontologist—an organizational and social scientist, designer, consultant, and inventor.

As a student of ontology, the study of reality, he developed some mastery of discourses—the conversations we have inherited that color our current understandings and beliefs about the world we work and live in.

His questioning of what we know gave rise to an invention that will powerfully alter how we relate to ourselves and others. This invention will dramatically alter our self-awareness of our actions, thinking, and interactions. What you are reading will introduce a new relationship with technology and a new world for us to live in. Organizations, governments, and families will gain a transparency and cohesiveness that only "simplicity" could possibly provide. You will appreciate the simplicity and obviousness of this new and novel orientation; however, it might seem too simple, too obvious. The author asks that you be patient, trust your experience, and doubt your conclusion that this sounds too simple to work. Please trust your instincts!

Sometimes when exploring or understanding something novel, there are moments of disorientation. If that occurs from time to time, it can be very powerful to simply follow the direction of the conversation and look for what experience or feeling is present, while giving up the need to understand or find something familiar. Allow your curiosity to foster your creativity and imagination.

Robert, why did you invent something to provide a new world to live in?

My work, until recently, has been limited to participating with a few organizations at any given time. Several of them experienced rapid growth and explosive profitability. These organizations afforded themselves opportunities to have an expanded awareness and coordination of operations, sales, and finances. I found myself interested in reinventing myself in such a way as to influence and contribute on a global level— shifting organizational and individual orientation.

I trust that we are ready, technologically. The technology to alter how we live and work is in place. I trust that humanity is ready for a radical break in reality and a dramatic shift in relating to ourselves and others in the world.

Why do you trust that humanity is ready?

Today, we are witnessing dissatisfaction with government. There are demonstrations worldwide. We are becoming increasingly aware. Some people are highly suspicious of politics and polarity. This isn't limited to nations. Discord and a silent dissonance seem to exist in the background of many organizations, communities, and even families. Some unsettlement arises in response to current issues. Some seems to be unresolvable, as it reappears time and again in our history.

We are more aware today than in any other time in history. From a humanitarian perspective, I think many of us are saddened by the atrocities and discord that dominate world news. We can examine and even understand our cultural diversity, but until we shift our relationship with ourselves, and until we gain an expanded awareness, we are trapped into repeating the atrocities of the past.

What do you mean about repeating the past?

One example is a fundamental plight for humanity. As we view our past, we see massacres strewn throughout history. Fairly recently (between 1500 CE and 1900 CE), history recorded the near extermination of indigenous people and members of secular groups. Atrocities took place in the Americas, Australia, Africa, and many other regions. In the last hundred years, genocidal conflicts include those of Armenia, Nazi Germany, Croatia, Bosnia, India, Bangladesh, Sri Lanka, Cambodia, Tibet, Haiti, Argentina, Nigeria, Ethiopia, and many other places.

Many are also disheartened by our failure to end hunger and provide adequate medical care to a majority of our world's population. Our lack of effective action doesn't necessarily reflect a lack of desire. I think we may be too distracted to act on our concerns for humanity. We are too hampered by our own inefficiencies to enable others to help themselves.

Robert, what will alter the future?

What is obvious to me, using conclusions from our past, is that only when we begin to relate much more effectively, become self-aware, and live inside a structure that enables us to act inside an awareness of all our interests and concerns can we begin to effectively engage in those matters that influence all of life. We must dwell within awareness: a lucidity of what more precisely might express, grow, and develop expressions that empower interests extending beyond our own.

I discovered three very old notions that hinder us. Until we recontextualize technology, we are compelled to continue to relate to technology as tools, the use of which take us out of relation with each of life's moments. Until we gain a full awareness of our communication, we unfortunately continue to perpetuate and blindly repeat the same patterns, deterring our commitments, relationships, growth, and development. Until our conversations occur inside of a structure for fulfillment, we continue in our life, work, and career, essentially missing opportunities to live a fulfilled and committed life.

I believe that somewhere deep within, we all recognize that the time has come for a new world. I trust this invention and book to open up such a world—to immerse all in a new awareness.

Is this book just about an invention?

This book does introduce an invention. This invention will profoundly reorient us, expanding our relations and awareness.

Dwelling in a New World reconstructs communication, technology, and accounting. My intention beyond revealing a life-altering technology is that your life's experience and understanding are profoundly altered.

By reinventing communication and our understanding of existence, you will gain an awareness of a direct connection with affinity, harmony, happiness, and serenity. The opportunity is for our relationships to become simple and natural.

This book serves as an opportunity to explore what we have taken for granted. Resignation and cynicism could be conditions of our current relationship with advertising, "new" products, and technology. Often we are thrilled by gimmickry and what is cool, but our hopes and aspirations are easily dashed by the crowded technological surroundings that we find ourselves living in.

To keep this introduction and book relatively brief, there are many directions that we will not investigate. This book will illustrate a shift for organizations, families, and individuals from linear process happenings to a (nonlinear and nonprocess) structural environment.

The nature of the inquiries and deliberations that follow serve to illustrate a surprisingly strong desire for living life. This book will endeavor to explore our desire, capacity for expanding awareness, and the invention permitting us to dwell in a new world. The book (and invention) also takes into consideration our desires.

- ✓ We are interested in specific areas of life (while suggesting that there are basic areas of life which appear to be consistently essential to us—areas we desire to have awareness and control of).

- ✓ We want to have people and resources that we can rely on readily available as support for these specific interests or concerns.

- ✓ We wish to gain simple structures to organize our interests (activities and developments that currently occur, will occur, or reoccur). We aspire to simplify life.

- ✓ We want to engage in interactions that are important (impacting our interests and concerns).

- ✓ We also desire a vast increase in awareness, ease, and effectiveness.

How will this invention satisfy our desires?

I will give you a sense of the invention, but don't expect to have a clear view of it until you have a good grasp on the basic new orientation that this invention supports. This will become apparent in part 4, after you get a sense of a new foundation for three different aspects of life, which together give us the substance for what constitutes our lives.

Before you can begin to explore this fully, you must engage in questions about how we are currently organized—the numerous applications and devices we use, reoccurring issues of miscommunication, and a life full of interruptions. At times, we find ourselves within a harried experience of fulfilling commitments. There are times when engaging in conversations that we experience disappointment of the outcomes of those interactions.

I ask you to relook at three different means of how and in which way we relate to life—what occurs as reality. I suggest that by redefining three aspects of our understanding and relatedness, we simplify life tremendously and enormously improve vision and self-awareness. I suggest that we desire happiness, satisfaction, effectiveness, relatedness, and awareness. This invention enables an environment that fulfills these desires, while the book, beginning in the first chapter, opens up a powerful capacity to see ourselves.

The status quo

What are these three different aspects of reality?

We inherited three very old paradigms. These three domains provide us with a day-to-day reality. These domains are communication, accounting, and technology.

The first aspect is communication. Exploring this aspect historically, it appears to exist today in very much the same way it did thousands of years ago. We have inherited that communication takes place when

- ✓ a speaker relays information or messages to a listener;

- ✓ performers perform for an audience; or

- ✓ a sender transmits something to a receiver.

Long ago, we used horns, drums, and smoke signals to communicate over distances. Distances were later traversed by couriers, telegraphs, and more recently, telephones, televisions, computers, and smartphones.

It is a rapid and sometimes not-so-rapid switching of the roles of speaker and listener that we normally refer to as a conversation. It might be obvious that, each moment, the listener's state of mind, his or her thinking or understanding at that moment, might influence the "information" heard. What may be less obvious is that the speaker's emotional state and thinking influence is what gets expressed. We are often aware of obstacles in relaying and capturing information. There are times when we become annoyed and disjointed as we rely on various methods, such as telephone conversations, e-mails, letters (mail), and texting.

The second aspect is accounting. Accounting has changed from the days of sea merchants, and we deal with very fast transactions today, yet the foundation of dual entry accounting (a linear and process-burdened double-entry bookkeeping system of debits and credits, first codified by Luca Pacioli over five-hundred years ago) has only been altered to meet current issues.

Organizations worldwide spend over a trillion dollars per year[1] and enormous costs in resources and time to anticipate or get a sense of current and future financial conditions. These resources include personnel, systems, and processes. Modern provisions are quite creative and expensive, with sophisticated resource-laden systems and dashboards—yet surprises and unexpected costs provide perils for even the best executives of well-run organizations.

Most accounting, such as financial, executive, customer, employee, and vendor relations management, is separate from interactions. Operations, sales, and other divisions usually partake in separate processes using different systems. Our telephone conversations, for example, can require many activities and interactions just to capture what was speculated, decided, or altered.

The third aspect is technology. For hundreds of thousands of years, technology has changed, but our relation to it has not. We still relate to technology as the tools, utensils, weapons, and now systems and applications that we use. The introduction of devices and robots just expands possible usage, which replicates and displaces human functionality.

While technological tools seem to benefit us, of those who have experienced multiple generations of data processing, computing, and networks—especially now that devices have become mobile—a few are disturbed by the changes they see. There appears to be less regard for our physical environment and greater involvement with our virtual world. While it seems that we have merely expanded our reach, our presence to life itself is compromised.

Even though recent thinking has provided some movement, we yet find ourselves living in a world continually distracted by communications.

We stop to record transactions or devote time and effort to organizing ourselves as we attempt to assemble everything we need for a process-laden project or to enter an address for our GPS systems, to illustrate a few examples. Up until this moment, we have seen no choice but to live most days as one interruption after another, with recurring distractions that interfere with our plans. We have systems that we need to maintain, processes to repeat, in order to organize ourselves.

How can we change these three different aspects of reality?

We have been altering these three domains for hundreds and thousands of years. There comes a time in history where we can no longer continue to change an existing model; we must reinvent or reconstruct it.[2]

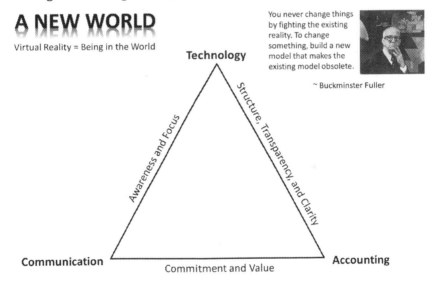

This book offers and introduces three new realms and an invention that brings a new way to live into existence. This shift isn't just a technological or an innovative shift; it is a life-altering and organizational, reconstructive shift.

What do you mean by technological or innovative shift?

Throughout history, innovations and inventions provided new tools, instruments, weapons, and so forth. A few examples are

- ✓ the printing press displacing scribes while expanding a capacity for learning;

- ✓ the computer (displacing manual instruments like the typewriter); and

- ✓ the mobile phone (smartphone) and tablets.

These innovations expanded our capacity to collectively interface with books, devices, tools, and applications (information).

In the last fifty years, computerization has moved from

- ✓ text to media;

- ✓ data processing to microprocessing;

- ✓ stations to networks (cloud); and

- ✓ from connectivity to browsers (mobile) to collaboration, with social networks as the most recent example.

The result of these innovations provides us with an expanded capacity for connectivity, research, and information, while at the same time causing a loss of emotional relatedness, productivity, and peace. It is said that we sometimes engage in activities resulting in hidden addictions. The way we are technologically organized often occurs as disruptive and unproductive.

You said something about technology today distracting or interrupting our lives. Could you be more explicit about this?

Yes, I was saying that we live in a certain condition where our personal, family, and work interests and concerns suffer from moments when our lack of a simple structure leads us to experience a world where demands come from many directions. We are bombarded by situations where we experience stress and are possibly overwhelmed. Interests, or what seems important, give us impetus to develop relationships, partnerships, and organizations. These arrangements provide for accomplishments (fulfilled plans), but given our current means of dealing with what is important, we also experience disappointments, disagreements, and breakdowns.

These breakdowns occur as interruptions to our plans and commitments. I suggest that we explore these occurrences and consider the possibility that they occur in an unintentional manner. Our lack of structure and order provides repetitive instances of

- ✓ distraction;

- ✓ interruption;

- ✓ lack of specific awareness;

- ✓ ineffective communication;

- ✓ conflict and avoidance;

- ✓ lack of ingenuity, reinforcing the status quo; and

- ✓ personal and organizational fragmentation.

It might be useful to provide an example. What might this look like?

What does a typical morning look like? Here is a short fictional story to illustrate how our current relationship with technology occurs:

> Bill and Alice will be celebrating their tenth wedding anniversary later this week. Bill is a marketing executive for a media advertising company, with a promotion to vice president last month. Alice has a pixie, petite appearance, natural grace, and a smile that can melt the deepest despair. She is perfectly suited to being the director of human resources for one of the top talent agencies in the city. Jake, seven, and Kate, five, rarely wake her up or disrupt her sleep. However, last night at 1:30 a.m., Kate woke Alice after having a terrible nightmare. Alice's smile quickly soothed Kate, and within ten minutes, Kate was fast asleep. Five minutes later, so was Alice.
>
> The alarm at 6:50 a.m. startles Alice as a stream of light enters into their bedroom. She feels the cool air on her face. Ugh, it is time to get up. She automatically silences

the alarm, gently touching Bill's arm as she does every morning. Still not thinking clearly and having only a vague awareness of the dream that was interrupted, she feels the carpet under her feet as she heads to the bathroom, not really conscious yet. Finding the light switch, the familiar bathroom comes into view. The simple rituals—washing the sleep from her face and eyes, brushing her teeth, and other hygienic activities—begin, though she's not fully awake. Rising to meet a new day, some sense of consciousness has begun.

These earliest moments are habitual, but she starts to diverge from prior days as she begins to think about the current day. While the plans for today start to become present, she continues with reoccurring patterns. In the kitchen, she starts breakfast, coffee; she wakes the kids and her spouse, and then she heads to her closet to get dressed. After adding just a little lipstick and color to her eyes, she heads back to the kitchen. She hears cartoons from the TV in the playroom and the news from the study, and back in the kitchen, she flips the bacon, turns up the burner slightly, jots down notes about what needs to happen today, pauses to remove the bacon, turns down the burner, and cracks open a half dozen eggs after draining the grease from the pan, yelling, "Breakfast is ready," as she puts the eggs on plates on the counter. The kids get their plates, and Bill gets the napkins, forks, and then coffee for the two of them, while Alice pours the orange juice. Everyone sits down to eat, a simple conversation begins, and the telephone rings.

Bill answers the phone; it is Alice's boss. He gives Alice the phone. John, Alice's boss, makes a request of Alice, which completely changes the plans she jotted down just a few minutes ago. She makes a second call and leaves a message, hangs up, and sits down. The eggs are cold, and everyone takes their plates to the sink. Alone, she quickly gobbles up the remaining breakfast, takes the plate to the sink, runs water over the plates, and leaves them in the sink. She is in a rush now. She encourages the kids to put their coats on and gather their books. Bill has already left. She opens the garage door and straps the kids into their car seats, and this time, her cell phone rings.

As she answers the phone, she remembers that she left

her list in the kitchen. She gets the list, picks up the dry cleaning receipt, sets the alarm, opens and locks the door to the garage, starts the car, and backs out of the garage. She is on the way to the preschool, and her phone rings a third time. This time, Debbie, the person on the other end of the call, sounds upset and needs her help. Alice gets to the preschool, watches Kate patiently as she walks into school, and starts off toward Jake's elementary school. She calls Bill, who abruptly says he can't talk. She arrives at school, and Jake and some other kids walk to the door. As she is watching, she texts Bill, "Can't pick up the cleaning. Can you?" She heads to a local shop before work to pick something up that's needed at work, and the phone rings yet again. It is 7:47 a.m., and Alice's day is only fifty-seven minutes old.

[To save time, let's move ahead two hours.] Alice has dealt with a few issues already—thirteen telephone conversations and five texts later. It is 9:51 a.m. She hasn't even begun a two-hour project which is due at 11:00 a.m. She sees that there are 127 e-mails waiting for her, but she can't bother with that. Her spouse texts her, "I can't get the laundry. I am swamped. Can you get it during lunch?" The office phone rings. Tom asks if they can move up the meeting for the project to 10:45 a.m. Alice's mind responds, *No!* Aloud, Alice asks, "Can we move it up past lunch—maybe 3:00 p.m.?" Tom says he will get back to Alice and hangs up. Her cell phone rings. It's Bill, her spouse, calling upset. She tells him she will get the laundry during lunch, somehow. The phone rings again. Alice discovers that the meeting can't be pushed back; the project she is responsible for is due in fifty-five minutes. It's back on for 11:00 a.m. She says that she doesn't have time to get it done by then, and she is told there is no option.

It is 10:07 a.m. Her cell phone rings, and she ignores it, opening up Excel and Microsoft Word. In Excel, she looks for the folder where the project is contained. She opens the folder, scrolls down to find the worksheet, and doesn't remember what Tom titled it. She opens up her browser, goes to her e-mail, and looks for the folder for this project. Not finding the e-mail, she does a search for "adjusted process," and two e-mails appear. She clicks on the first, dated March 9. This is the one she needs. She returns to Excel, clicks on the date to sort the files and scrolls down to March 9. There it is. She opens up the file. She switches to MS Word, locates the same folder, sorts by date, and opens up the Word document as the phone rings. Her cell phone is also flashing, informing her that there is a message waiting. She ignores the phone and starts reading the document.

Interrupting her reading, she glances away and moves the Excel document to her second screen. She looks at the tabs at the bottom for the most recent version of the project, and an employee walks into her office saying that he needs to talk. She glances at her watch, sees that it's 10:21 a.m., and says sternly, "I can't now. Get back with me after lunch." Mike, the employee, says, as politely as possible, that this can't wait—could they speak during lunch?—at which point she remembers that the dry cleaning needs to be picked up and ...

We could continue on with this fabricated story, but in these last few minutes, we have gotten a sense of how thirty minutes might occur in a typical morning at work. If we had continued with all the details—the writing, rushed preparation, and so on—we would find out that Alice was late to the project meeting, ill prepared, and that, subsequently, she and John were left fighting to keep the client from going somewhere else, as the coordination of personnel within the client's proposal was too sketchy. John, her boss, talked to her during lunch. The laundry never got picked up, and Mike, never meeting with Alice and her infectious smile, quit!

I can really identify with this story.

Unfortunately, we can all identify with this scene and, in many cases, it might be way too simplified. Many of us have to deal with numerous computer applications like Microsoft Office (Word, Excel, and PowerPoint), systems like CRM (customer relations management), PM (MS Project as an example), and Visio, e-mail systems with calendars, group calendars, tasks, meetings/events, and chatting, post-it notes, notepads, telephone conversations, face-to-face meetings, and conferences ... We could go on and on. This list only deals with work and the activities there. If we have laptops, we get to take these systems and applications home with us—and even on trips.

What happens when we open our e-mail and there are 137 unread messages waiting for us? These e-mails range from junk to pressing issues pertaining to a number of different accountabilities. We are bombarded with this influx of communications, diverse in interest and concern, yet arriving in a way in which they appear simultaneously. Dealing with this morass can be challenging. What about the dozen additional phone messages that are waiting for us? These messages also range in importance and relate to many different issues, but again, we deal with them linearly, one at a time, putting some off and tackling others.

What of our interests and concerns outside of work? They don't stop. People rely on us—our friends, spouses, children, and teachers. If we belong to organizations or volunteer from time to time, where does this fit into our day? And let's not forget about our workouts three times a week, our visits to the doctor, the children's soccer games, and calling our siblings, parents, cousins, or grandmothers. Then there are everyday things, from eating and meeting with friends to writing our representative about the educational bill that we want passed.

Almost everything, every important interaction, seems to result in processes once the conversation is completed. We are left to schedule something in our calendars, look up phone numbers or e-mail addresses. Sometimes when something in our schedules resurfaces, we have to stop and figure out what purpose or importance the entries represent.

There is an alternative that eliminates processes altogether. It eliminates opening folders and looking for e-mails, documents, phone numbers, and so forth.

Radical shift

What is this alternative?

The alternative is a drastic and radical shift. By reconstructing our technological orientation, our virtual and physical world will exist in unison. Our virtual environment will simply support us in the world. It will bring us powerfully into the world, alleviating distractions rather than creating them.

A common problem occurs when new foundations for understanding are born. Historically, these shifts seem to offer little relief, as the older model or paradigm continues to assert itself. We need support.

I assert that we will be immersed in a new world, where technology, communication, and accounting exist in a very different way than they presently do. Unlike the printing press, which allowed numerous books to enter the world, this invention will create a new orientation, a technological environment that eliminates processes by providing a means to create structures that empower our current activities and our future interests and concerns as we relate and interact.

This sounds like we are going to be transported into a new world.

First, remember I am introducing this invention without providing you with a powerful orientation. I anticipate you will experience a lack of clarity with this first look of a virtual environment that supports your physical activities. Unlike gaining access to this environment experientially through the invention itself, we are only discussing this reorientation.

The illustration above gives you an intellectual understanding but without the powerful background that will allow you passage into a new world.

Yes, a new world whose existence requires a new construction of three ancient domains. In our new world:

Technology no longer exists as tools and applications (processes and procedures) and does not occur to be device-centric. Technology becomes a world in which you "dwell." Wherever you are, whatever you are doing, it supports what you are currently interested in, what you want, and where you are going.

Communication ceases to be linear. It does not occur as the mass of separate, linear discussions, which gains in complexity and volume over time. Communication is no longer limited to language and doesn't manifest itself in constant interruptions. You will not only live in a "you-centric" orientation but an interest-centric environment too—where each shift in interest provides an environment (including people and resources) that supports what is important. This dwelling allows you the comfort of having what you need, when you need it, as you shift into planning an event or going out to eat. Communication expands from a single dimension into three, providing depth and self-awareness never experienced before, with a level of effectiveness (and support) that expands over a lifetime.

Accounting no longer occurs as entry processes or a series of actions requiring a clerk. There is no need to account for and enter something into an accounting system. Accounting occurs within the interactions themselves, supporting your every intention, thought, and arrangement, as this environment provides the very structure for our suppositions, plans, and commitments. As plans and commitments are made, change, or become completed, the past, present, and "current future" are directly related to and accounted for.

What does "current future" mean?

This might only appear as a concept as I attempt to explain it. You and I will be living in a world where the future, as currently planned, is very visible and precise. Our current future changes each moment as we discover more effective ways to plan or act, or as people with whom we are coordinating accountabilities change their plans, and the old future gives way to a new one. The implications for governments, organizations, and families are profound because each person and his or her actions, interactions, agreements, commitments, and speculations provide vivid futures as his or her decisions, thinking, and negotiations influence finance, resources, and strategies. This new environment is much more powerful than is possible to describe; however, you are starting to get the gist of something extremely powerful but immensely simple at the same time.

It sounds like you are talking about a sci-fi or futuristic reality.

The first thing that you must confront is how foreign or alien this new world sounds. Can you imagine how someone describing an airplane might have been perceived a little more than a century ago, when ships, trains, horses, and wagons were the primary available means of transportation? This invention will bring forth a new world unlike anything ever experienced before. While I anticipate many new future developments spawning from this reorientation, we have the technological means today to powerfully enter into this new existence. No new technological functionality needs to be invented. This invention simply redesigns our current capabilities from a linear application and process-driven arrangement to a virtual dwelling or structural environment.

Within this environment, governments, organizations, and families will naturally reinvent themselves. Structures are innate for humanity, and intentions (interests and concerns) are the simplest building blocks for designing any sized organization, as well as each of our personal desires.

One thing that some futuristic, sci-fi movies and shows imply is that an environment of acute awareness leads to a world where humanity becomes peaceful, where food and medical care are available for everyone. It doesn't surprise me that what I express and envision for humanity appears like science fiction. It doesn't shock me that considering or discussing humanity's entrance into a new paradigm has been and may be met at times with disbelief, resistance, and even antagonism.

You sound self-assured. How do we know this isn't just your imagination?

It is my intention that, by the end of this book, you are self-assured. Imagination is often mistrusted and dismissed. Our culture stipulates that being grounded in facts is important. There is an old cliché—"I'm from Missouri!"—that is not limited to residents in a particular state. Albert Einstein said, "Imagination is more important than knowledge. For knowledge is limited, whereas imagination embraces the entire world, stimulating progress, giving birth to evolution."[3] It requires a willingness to remain curious in order to see an imaginative notion as an interesting marvel. At times, we appreciate imagination and our experiences, while other times we dismiss them.

As we begin reinventing communication, in the first chapter, we will explore imagination and experience. For the moment, let me suggest that imagination opens up a universe for new thinking. The onus for me will be to provide obviousness to this new thinking, but the invention itself will bring us (structurally) into this new world.

Can you say anything that will allow my imagination more clarity?

Regarding the invention, for now, just think about living in a world where each moment you interact with a supportive "virtual entity" or companion that alters and relates to what's important. It occurs as a technological space that is interest-centric—a communicative, interactive, accounting structure that will take some elucidating to get a better picture.

A relevant analogy might be the rooms of a home, which support different interests that take place in each location. The bathroom and kitchen, more specifically, provide different orientation, design, and purpose, as the interests and concerns related to these two rooms differ. We could say that these rooms provide a different space to live in. Living in this new world, a technological, virtual space supports the current interest or concern we are dealing with. As our interests change, so does the virtual environment, moving us into a new virtual room that nurtures our interest or concern.

How about a functional diagram of this virtual companion? Don't think about it as a device or something you need to carry; I will get to that next. For now, know that devices are no longer the issue; you are. This diagram gives you a functional way of relating to what is important to you, nothing more. Think of something specific that is important to you and look at this diagram. The people and resources and the interaction that involve this specific interest or concern are all that are present.

Interests and Concerns	Interactions and Activities	Invention Diagram

Interests and Concerns

Important Areas of Life

Organizational Design and Accountabilities

Resources and Relationships

People, Organizations, Resources, etc.

Support and Structure

Lists, templates, tools, utilities, systems, gadgets, accounting, schedules, calendars, controls, settings, etc.

Interactions and Activities

An environment where collaboration, research, speculation, planning, negotiation, sales, etc. occurs.

All activities, media, communication, entertainment, reading, and so on either occurs or is supported in a virtual world.

A virtual world that parallels and alters as you in your physical world move from one interest or activity to another.

Real-time Feedback of Effectiveness

A Novel Communication Breakthrough

Invention Diagram

Wherever you might be each moment (at home, in the car, at work, in stores, and so forth), the nearest display(s) provides an interactive, virtual, interest-centric environment for you, constructed to manage nonlinear interactions, media, drawing, diagrams, forms, directions, and audiovisual discussions on an electronic whiteboard.

These interactions are supported by interactivity-accounting (scheduling) capabilities by way of interest-specific resources, systems, templates, tools, and applications. Your environment is configured to perform, support, and empower effective emotional states, interests, activities, and communications for you and the organization represented by this current interest.

Interactivity accounting provides current, present, and future allocations of resources and finances. Templates and several structures (systems) support autonomous yet coordinated activities and interactions for you and the organization.

An illustration follows, of what a snap shot of a screen might look like.

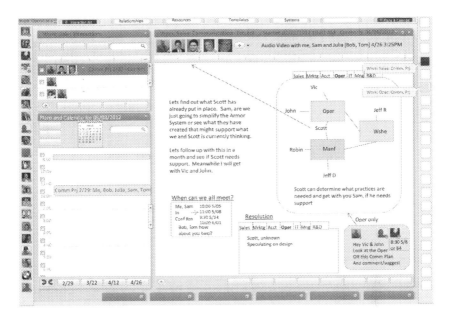

A more descriptive clarification of the invention

The invention (patents pending) proposes a new field of technology. It reconstructs existing technologies.

The invention introduces a new field of communication that provides an expansive awareness that empowers effective interactions. Current and reengineered technology supports this new communication paradigm.

The invention announces a novel field of accounting where the current future is visible and directly accessible. The widespread use of existing technology that supports us in this new environment is also proposed.

The invention provides a background which governs a migration from the use of many diverse, screened, device-centric contraptions to an entity-centric, uniform environment that supports interests and concerns. It provides specific support for current actions, focuses, and interactions.

The patent-pending involvement completely reorganizes technology as a space instead of an application. It lays out a powerful, self-aware communication and accounting (future transparency) model. It spells out a methodology with precision but offers flexibility of design suited to each person or organization so they may create their own virtual homes—with rooms, furniture, and all the "stuff" like people and resources to support each interest in a structural way. This is an orientation that creatively destroys process-driven designs.

This sounds like something for companies. How might this benefit everyone?

This new world—a world where the way we relate to ourselves and others—expands rapidly as the virtual companion (the invention) empowers a natural capacity to reinvent organizations. This new technological space simplifies organizations, supports mentoring (awareness and control) and effective interactions (selling, negotiation, research, making agreements, and so forth), and gives a clear view of the current future for resources and finances. Manuals, procedures, and processes will disappear, as structures for design provide a much more potent environment and functionality.

This simple structure also benefits families. Essentially, families are small organizations, at times lacking in structural support and awareness. There are many specific interests and concerns pertaining to families. Finances might be the most obvious interest, but there are a great number, the coordination of which supports the family unit.

As individuals, we are engaged by organizations, and our relationships are enhanced. As we move through our everyday activities, what is important to us becomes structurally designed to support us, raise awareness, and help coordinate our lives in a simple and peaceful way.

This still sounds vague, but I am starting to realize how different things will be.

I am just beginning to describe and illustrate what will likely be available for us later in 2012. No matter how much I describe the new world provided by this invention, until we are living in it, we can only talk about it, using our imagination to discover a completely new and powerful way to relate in the world. By the time you complete this book, you will have a much clearer notion of this new world we will be dwelling in. The new communication shift might provide a more immediate effect. This is only the introduction. I think it might be of benefit to explore the three domains more fully, giving you further foundation for this new environment.

The next three parts are critical for enabling you to climb into the last part, part 4, "The Invention." In these parts, you will gain further awareness of the three new realms that we will be transported into as we dwell in a new world. Part 1, "A New World vs. Our Current World," opens up a new context for communication and a technical means for gaining an unprecedented capacity for self-awareness and effectiveness. Part 2, "A New Way to Relate to Technology," gives you a sense of the radical shift necessary for us to live in the envisioned world. We will soon exist in a technological space that enhances our waking moments beyond our comprehension. Part 3, "Interactivity Accounting: Powerful Awareness, Scheme, and Structure," leads us into an exquisitely simple structure that provides individuals and organizations with a current view of the future that serves as a background for our relations in the present moment. Subsequent actions—processes—simply disappear.

This book is very short, requiring little time to read. There are three chapters for each part except part 4, where notes occur in the back of the book. Most of you might read the first chapter for each part (chapters 1, 4, 7, and 10) and simply bypass the notes in the back, as these four chapters will provide a basic, coherent understanding. In terms of reading, you are now a quarter of the way into the basic book (and one-sixth of the complete book if you elect to read all ten chapters and notes).

For those who are prone to look under the hood or want a more specific philosophical or technical comprehension, the second and third chapters develop the primary chapter in each part more fully. These two additional chapters and the notes in part 4 might appear drier or more technical in some ways, or they might expand or dwell in more abstract or possibly rigorous thinking. Utilize the notes and two additional chapters in each part of the book as you desire. Their utility may provide clarity. If you are not technically or philosophically inclined, reading these chapters and notes might raise new questions. (A viable way to read this book, a quasi-nonlinear approach, is to first read chapters 1, 4, 7 and 10 and then read chapters 2, 3, 5, 6, 8, 9 and 11.)

There were many people who provoked my imagination and contributed to the reconstruction of three ancient domains. Let this book bring you to some of my current conclusions, a resting place for my explorations. I may write books in the future that provide the historical, empirical, and organizational research that led to these paradigm shifts. These future books might appeal to scientists, historians, anthropologists, philosophers, or theologians more so than this book, but this book, especially the primary chapter for each part, is for you. It will require imagination to bring us into this new paradigm, necessitating that our imaginations catch a glimpse and keep hold for several months, until the invention is fully developed and brings us into this new world.

1. "IDC Predicts a 1.8 Trillion dollar IT Industry in 2012," IDC, last modified December, 1, 2011, accessed May 22, 2012, http://www.idc.com/getdoc.jsp?containerId=prUS23177411

2. "The Buckminister Fuller Challenge," Buckminister Fuller, Buckminister Fuller Institute, http://challenge.bfi.org/movie

3. Albert Einstein and George B. Shaw, *Einstein on Cosmic Religion and Other Opinions and Aphorisms* (Mineola, NY: Dover Publications, 2009), 97.

Part 1
A New World vs. Our
Current World

Chapter 1
What Is This New Communication?

Education, conclusions, and inheritance

How did imagination open up a new way of thinking?

In my studies and research, I came upon many brilliant people and the theories that they expressed:

- Albert Einstein—Theory of relativity

- Fernando Flores—Ontological design

- Humberto Maturana— Neurobiology (biology of cognition and love)

- Martin Heidegger—Reality (ontology—phenomenology)

- Christopher Alexander and Louis Kahn—Architecture (spatial design and language)

- Melanie Klein and Wilfred Bion—Object relations theory (psychology of discernment)

- Brian Regnier—Exploration

- Michel Foucault—Anthropology

I have noted several of these and other figures whose wisdom I build on and whose conclusions I question throughout this book. I indicate their contributions and my resulting explorations to different degrees in different chapters.

3

Numerous others have influenced my fascination with several schools of thought. The various theories and rich intellectual influence of each of these people lead me to appreciate that understanding is only a temporary conclusion. Only by questioning both my conclusions and their theories could my imagination take me beyond present-day thinking.

While this book and invention reshape how we relate to ourselves and others, I suspect that this new relationship with and way of living in the world will evolve. This break with our inheritance could stimulate thinking and eventually open up these paradigms for even simpler models in the future. Will this occur in a dozen years, a hundred years, or a thousand years? I don't know. Regardless, this new environment opens up a rich new way of viewing how we know what we know.

The first thing that seemed obvious to me—and where I elected to pause as a place to look from in developing a new model—is that we evidently have an awareness of what is internal and external to ourselves. All living creatures seem to be conscious of what is internal and external to the body or membrane that separates life from the environment. This is a basic notion: internal and external. Regardless of our sensory perception and all life's senses, this is a basic dualistic aspect of life.

Experience

What do you mean by the term "dualistic"—internal and external?

For human beings, there are numerous references to the concepts of internal and external, and they span many disciplines. A simple analogy is seen in comparing the technology of a biologist with that of an astronomer. When I was very young, I became curious about the lenses of microscopes and telescopes. I could view very small, nearby objects by looking through the microscope, while I could view large, faraway objects by looking through the telescope. This example, early in life, exposed me to a dual utilization of lenses as they occurred in two different fields of science.

From a biological perspective, life appears to have some kind of barrier that separates what is internal from what is external. For cells, it is a cellular membrane. For human beings and more advanced life, different surfaces, such as dermal (skin), ocular (visual), aural (hearing), and olfactory (smelling) systems coordinate to provide this barrier.

Living things in general and human beings specifically appear to have senses (such as touch and taste) and sensors (sensory organs, glands, and the like) that reveal the internal and external receptivity, accessibility, and connectivity for life. Life seems to be a closed system that is influenced by its environment.[1] Physics has several dualistic explanations that provide completely different models for life and existence.

Moving beyond biology and physical attributes, we find many different theoretical models that portend to explain life. These scientific models require a proposition and experimental evidence to support relevant theories and are later replaced by more current and up-to-date models. Such is the case with current neurobiological theories about the human nervous system and brain. It is doubtless that, in time, the current "information storage" type of modeling (which resembles the current computer technology thinking) will give way to new theories.

Right now, I am interested in exploring something simple. Regardless of propositions and experiments, I am interested in what precedes the establishment of new theoretical models. I suspect that what might initially appear to be experientially derived is simply a conclusion that the evidence supports.

Regardless of the frame of reference, life occurs in and is attentive and responsive to two realms—those occurring as internal and external.

What does internal/external have to do with this new model?

Our senses—hearing, seeing, tasting, smelling, and feeling—seem to give us, *in part*, the capacity to experience. What we experience might be internal or external. Experience has an immediate and momentary quality. Just as a hand has two sides, experience can occur as an internal presence while the second, active component is expressive in nature. Examples of the expressive component are listening, speaking, and touching.

Another way to view a dual nature with internal/external and experience/expression is to think of "internal feelings" as experiential—and expressions, actions, and the like as expressive, external happenings. We seem to be able to experience two other worlds also. One world defies understanding, and in the other world, understanding takes place. We can begin to explore this if you are ready. First, let me summarize.

5

This notion of internal/external plays on the dual nature of experience and actions. It provides the momentary experience and activities or expressions that occur in a moment of now. These senses (which seem to be much richer than the "five senses" usually mentioned) provide two sides: internal and external, experiences and expressions. Actions and experiences are poignant by nature. They have active, momentary aspects that are emotionally attuned to aggressiveness when thought of in terms of communication. Communication has a connective aspect to it that we will soon delve into. The new model of communication has three components. This first component is experience, action, or aggression.

Reality

I don't view communication as experience, action, or aggression. Expression is closer to what I think about when I consider what communication is.

Can you say more about how communication is normally viewed?

The ancient model that we have inherited involves a speaker, a message (or information), and a listener. Historically, communication occurred orally, signally, pictorially, and later in print. Some of the earliest signal methods were smoke signals, horns, and drums. These evolved into the symbolic and pictorial—cuneiform, hieroglyphics, Hebrew, Chinese—and eventually the alphabetic languages arose with Russian, Greek, and Latin. In more modern times, humanity has used the telegraph, telephone, digital methods, Internet browsers, audio/visual equipment, and the like. Numerous medias like radio and television are now in use.

> These are the current (Merriam-Webster) definitions of *communication* (noun):
>
> **1:** an act or instance of transmitting
> **2 a :** *information* transmitted or conveyed
> **b :** a verbal or written message
> **3 a :** a *process* by which information is exchanged between individuals through a common *system* of symbols, signs, or behavior <the function of pheromones in insect communication>; *also :* exchange of *information*

b : personal rapport <a lack of *communication* between old and young persons>
4 *plural*
a : a *system* (as of telephones) for transmitting or exchanging *information*
b : a *system* of routes for moving troops, supplies, and vehicles
c : personnel engaged in transmitting or exchanging information
5 *plural but sing or plural in constr*
a : a technique for expressing ideas effectively (as in speech)
b : the technology of the transmission of information (as by print or telecommunication)

A common-sense understanding is that communication is an activity of transmitting information. Communication requires senders, messages, and receivers that need not be present or aware of the sender's activity to communicate at the time of communication. Communication might occur across vast distances. Communication requires a common communicative understanding. Once the receiver has understood the sender, the process is complete.

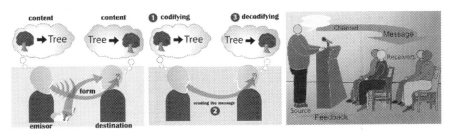

As illustrated, the communication of ancient days, when Aristotle captured the speaker → message → listener model, has been adjusted and transformed in today's times. Our current understanding of communication conserves this model, which contends that all information—whether verbal, nonverbal, or through signs, diagrams, pictures, media, and the like—is relayed or initiated by some sort of language. Languages might align more readily with your thinking of expressions as being perceived in contrast to actions, experience, and aggression.

This current view of communication seems real. Why mess with it?

I could retort with Albert Einstein's view of the physical world, reality, and thinking[2], but I won't right now. I think that the ineffectiveness of communication is often overlooked. We ignore any faults so that communication can appear rational and explainable, but the cost of its ineffectiveness takes a financial and inhumane toll. When we think of messages as expressions, or language expressed, we must take for granted three notions: the listener, speaker, and information. For this model to remain intact, we cannot explore communication beyond the activity of language. For instance, we're unable to examine the interruptions that take place during this activity.

Looking at this model, communication is portrayed as information and as an expression (in which "language" is spoken and listened). At least two components are quickly thrown together, in which what is central or relayed is understood as information in language.

My re-creation of communication overpowers some of the most brilliant contemporary philosophic thinkers. Ludwig Wittgenstein said, "Now I am tempted to say that the right expression in language for the miracle of the existence of the world, though it is not any proposition in language, is the existence of language itself."[3] While I do not oppose the importance or capabilities of language—"Language is the house of being. In its home human beings dwell."[4]—I have come to the realization that this way of thinking is entrenched in an ancient paradigm.

The nature of your lead-up for your question is very revealing. "This current view of communication seems real." A very narrow view ties communication to what is real, to reality.

What else is there but reality?

Imagination might exist outside of reality, but—besides imagination—I see a vast existence in which reality plays a very small part. Up until this moment, we have placed a huge amount of importance on reality. This makes perfect sense, as dictating what is important happens solely in this world of language, this world of reality. This possibility opens up a second aspect of communication.

The first aspect, experience/expression, is an internal/external phenomenon existing only in the current instance: now. Only a small portion of this aspect is considered in the current model of communication, and this is limited to mannerisms, speaking, and listening. This is the portion of experiences and expressions that we are conscious and aware of.

This second aspect of communication is strictly internal—founded in learning, understanding, and remembering. For humanity, this is the world of language, of thinking, of recognition and cognition. Reality exists within this aspect. While this domain of communication might seem to cover a huge expanse, it is the least prevalent aspect of communication.

How is that possible?

This idea, the assertion that thinking (or cognition) might be a tiny portion of our existence, seems irrational, doesn't it? If you consider thinking, this is where time exists. It is within this temporal framework, where we understand what we see. It is this framework where we are attentive to and conscious of our plan for the future. It provides us a totality of what we understand that exists. The past, present, and future appear in our thinking. When we tie together successive experiences, the motion of life comes into existence. The only way we can recognize an experience is to remember and recall it.

What is this notion that thinking comprises a very small percentage of life?

I think this is one of the most difficult ideas to deal with. This might be why we are so entrenched in our current model of communication. The whole experience of time, especially the past, seems to be the whole banana. For all of existence, everything that has ever happened exists in and occurred in the past. Until we grapple with the idea that time occurs in thought and not in the world, we cannot see any limitation to time.

Another conception of life that makes this notion sound foolish is that life is simply what we remember and think of it. What we want, what we believe, what we plan, and what we remember appear to be completely inclusive and comprehensive. Don't they?

> I prefer an attitude of humility corresponding to the weakness of our intellectual understanding of nature and

of our own being.
—Albert Einstein[5]

But what about all those moments we don't remember? What about all that we are not cognizant of or don't formulate into thoughts? How much of life is experienced and not remembered? How much is remembered and then forgotten? Do you sense something extremely vast beginning to open up? Even if we let go of all that we have forgotten, if we get present to each moment, what percentage of our experiences do you think we will actually think about or remember? (There is a vast world of experience that we are unconscious of, even though we are responsive to it.)

Forget about those hectic moments; what percentage of the quiet moments do we recall? I am asking all of these questions because I am interested in your gaining the experience that we only recall or are mentally conscious of a tiny percentage of what we experience. It is necessary to challenge the assumption that we are attending to any significant portion of our experience.

Another aspect deals not with what we recall but what we project as the future we want to live into. Daniel Gilbert brilliantly devotes a whole book to illustrating the futility of trying to design life for happiness and the futility of our many attempts and aspirations.[6]

Experiences aren't part of reality; they aren't rational. We can't predict or understand what makes us happy. We can't predict or understand what has us experience love either. Our thinking that some actions equate to love and other actions don't is a prescription for misery, disappointment, and loneliness.

A while ago, when I first asked about our current view of communication appearing so real, I asked you if there was anything else but reality. You said that imagination might exist outside of reality. Can you expand upon that now?

This is an odd discussion to have, but it is central to what makes humanity so remarkable. While language and cognition provide edges for our comprehension, our reality, there is fuzziness to these edges, a frontier for exploration. Our experiences (and a realm we haven't discussed yet) seem to foster this phenomenon, this ambiguous zone. Our capacity to unlearn, or let go, of the importance of understanding promotes this quasi reality that we might characterize as imagination. Imagination isn't always well thought of, and *a speculative space* for research might look more politically correct.

It seems that curiosity and inquisitiveness foster this ambiguous and even ubiquitous state, where thinking of thinking occurs as a freedom from constraints. It might require a diminishing of the importance of comprehension to reach into experience or an experiential realm. There is something sacred and reverent about curiosity or *wonder.*

We can think beyond thinking, we can experience that, we can imagine it, and we can wonder. Now, for us to remember this revelation, for us to express it, we must grapple with speaking it: formulating new words or connecting ideas and abstractions differently. In that grappling with it, however, we lose much of it, even as we invent new words, concepts, and terms to express it. It is this realization that has me trust experiences while distrusting conclusions; it is conclusions that are expressed as words.

Are you saying that thoughts aren't important?

No, I think that our thoughts are the very essence of what is important and have a powerful influence on our emotional state. I am just suggesting that our emotional experience has a very powerful influence on how or what we can think. The proverbial question arises: which comes first— experience or thought? Unlike the question about the chicken or the egg, this question has an answer.

Neither experience nor thought comes first. This opens up the third aspect of communication. This aspect can be the most elusive. What are you aware of so far?

I am aware of experience and expressions, which have internal and external characteristics. This aspect only occurs now. I am aware of where reality comes from. It seems that I think about things and ideas, and I remember and understand what occurs for me. My interests and sense of importance are a product of thinking. The past, present, and future appear as what I discern. I can see that continuity and motion appear as I discriminate among what is happening. I can understand that comprehension and discernment are part of an internal function called thinking.

Wonder

What is this elusive third aspect of communication?

If you recall a diagram in the introduction (illustrated again, below), one of the boxes is titled, "Real-time Feedback of Effectiveness: A Novel Communication Breakthrough." Our discussion will expand this premise. Much of humanity struggles to seek or find happiness, peacefulness, and love. For some of us, these states of existence seem elusive—and we "just can't seem to find the one" is a common expression.

Interests and Concerns Important Areas of Life Organizational Design and Accountabilities	**Interactions and Activities** An environment where collaboration, research, speculation, planning, negotiation, sales, etc. occurs. All activities, media, communication, entertainment, reading, and so on either occurs or is supported in a virtual world. A virtual world that parallels and alters as you in your physical world move from one interest or activity to another.	I n v e n t i o n
Resources and Relationships People, Organizations, Resources, etc.		D i a g r a m
Support and Structure Lists, templates, tools, utilities, systems, gadgets, accounting, schedules, calendars, controls, settings, etc.	**Real-time Feedback of Effectiveness** A Novel Communication Breakthrough	

Your patience with what might seem a philosophical exercise could reveal what drives us to obtain things and take actions, which may disappoint us over time. *One aspect of dwelling in a new world will give us feedback on what diminishes peace, affinity, and happiness, as opposed to searching or looking for it.* This invention supports us in recognizing the three levels of communication, each occurring on completely different planes from one other yet—like the different sections of an orchestra (woodwinds, strings, and brass as an example)—giving us music or what occurs as our lives.

The third aspect of communication has a timeless and nonpositional aspect to it. Coupling, connecting, and relating point to this third domain. Even though I am expressing in terms of understanding, this third aspect is further removed from understanding than experience is.

What do you mean by further removed?

Experiences happen in an instant. If we are conscious of or think about the experience, thinking about or remembering the experience is very distinct from the experience itself. Smelling a rose and remembering smelling a rose are different. One is an experience, while the other is a sense of the realness of something experienced.

In this way, experiences appear to become real as a memory or as some occurrence recognized as having happened in the world. The experience has passed, but the knowledge of the experience is present.

The timeless and formless third aspect can only be accessed (touched) through experience. It might occur as an experience of beauty, happiness, or stillness. The experience, however, is of a relationship to beauty, not something beautiful; it might be a relationship with peace but not peace itself.

This third aspect is always in the background. We are never cognizant of it directly, but we can be aware of our experience of the stillness, peacefulness, connectedness, love, and joy. In those moments when we are very peaceful and thoughts have disappeared, in the silence, we experience peace, joy, and love. We experience the depth and timelessness of this third aspect until our thoughts rush in, and then we experience our thoughts.

How can we expand our capacity to experience this third aspect?

Expanding our awareness increases our effectiveness. While I assert that attentiveness to these three levels is liberating, I think we need help or assistance, and I anticipate this benefit from the invention. Humanity has some basic interests that support and open up opportunities to experience this third aspect. A few of these interests are spirituality, friendship, play, and family.

What we are discussing now moves forward where Gilbert's *Stumbling on Happiness*[6] left off. We gain an expansive capacity to experience this timeless state in relationships where love is present or joy is prevalent. "Love" seems to encapsulate our experience of this third aspect, as love might be described as wonder, beauty, and joy.

Using the term "love" is confusing, as it seems like an internal feeling.

With love, given how we view the acts of relating and communicating, there appears to be much to theorize about and explain. These objective attempts to understand an emotional state create some mischief. Voltaire wrote, "Love is a canvas furnished by Nature and embroidered by imagination."[7] It's possible that the source of the confusion is that love doesn't appear to be an internal phenomenon. It seems to be a relating phenomenon, one that we can have experiences with. The experience of love can often seem internal, but not that which gives of the experience of love.

For instance, at times I get aggravated with my wife, Gayle. My irritation is experienced as a blend of feelings and thoughts. Even though these moments are not pleasant, and there is an aggressive expression internally and/or externally, a connectedness underlies these occurrences.

Happiness and love are misunderstood because they defy understanding. Emotions like empathy, compassion, attachment, longing, and well-wishing all piled under that one word—love—are a blend of experiences and thoughts. Our expressions seem to be on the other side of the hand from experiences. Experience and expression seem to exist, respectively, as internal and external happenings. This subject is rich enough to fill another book.

Communication defined with speaker, message, and listener is the paradigm we have inherited. As we attempt to see ourselves as rational beings, our discussion must certainly undermine this inheritance. Does this open anything up?

Does this mean love isn't something to receive, that it is about giving something?

I invite you to look at a relationship in which you never doubted that love was present. Over time, you recognized that there was always a deep experience of being connected. Even when you were furious with another person—or that person with you—the relationship was deeply seated.

You or the other person might not have been particularly nice at times, but your relationship was consistent. If you can recall this experience, I doubt there was anything tangible to receive; there was a steadfast connection. There may have been no sugar coating and no real fear of upsetting or displeasing each other but always a remaining, profound connection.

There is something that's not easily understandable about being intimately related. Relations could even appear hostile at times, but at a level where knowing doesn't matter, there is a space within the turmoil where a release occurs and this connection is present.

Can you come up with a simple example of this domain of love, peace, joy, and beauty?

Okay, let us consider that cognition, or reality, requires language. We require language to distinguish a flower as beautiful. Now let us think of this flower, you, and a newborn baby. The main difference between you and the baby is that the baby hasn't experienced as much of life, hasn't been involved with recurring patterns of language, and so cognition and distinctions haven't yet taken a strong hold. For you, there is a "flower" and you can experience a relationship with this flower; you experience beauty. For the baby, there is no flower, only the experience of being with beauty.

A new model

Where did you come up with this new communication model?

I read a lot of different ideas and theories. Then, I put everything aside and let my imagination explore.

Everyone has this balance of the measurable and unmeasurable in him: but a man like Einstein would be one like the poet who resists knowledge because he knows that if he were to talk about what he knows, he knows that it is only a miniscule part of what is yet to be known. Therefore, he does not trust knowledge but looks for that which can be order itself. He travels like a poet, the great distance, resisting knowledge; and then when he does get a smidgen of knowledge, already he reconstructs the universe ...
—Louis Kahn[8]

I took a little journey, looking at even the simplest living creatures, and imagined. What happened in the beginning? What happened next? A story of life developed from my journey.

In the beginning, there existed a connection between one undifferentiated thing and another. The connection between itself and its surroundings appeared to be enough; it was sufficient. The being was startled, then comforted, and sent off to quiet sleep almost immediately. There was a sense of belonging, a face, a breast, the experience of comfort, discomfort. The sensations were different—not fluid, contained, or warm. The environment was startling, and then everything seemed okay. Belonging, a face, a breast, comfort, discomfort, different sounds, different sights, different textures ... Although I had no way to recall what a baby experiences, I studied the observable, and I imagined.

Something profound in the animal kingdom, with mammals especially, is the almost constant suckling and caressing with a mother and her offspring. In this, we see something that needs no explanation. It is simple yet profoundly intense. For this baby and his or her first reoccurring relationship, there is a foundation, a connection—love. I surmised that love appears to be a natural phenomenon, and the basis of life from an emotional perspective is love.

The world and our internal sensations often disturb and interrupt this life and its foundation of love, or connectedness, which allows all other developments to occur. Love connects life to its surroundings and to all beings. In the beginning, there exists an undifferentiated love—curious, alert, and alive. But the sensing of these interruptions, these perturbations[9], introduces an altered experience, a differentiation. A response, a movement, something aggressive when compared to the background of loving peace, is aroused. But the calmness and silence—and peace and love—persist.

Love doesn't need much of a description; it remains as a timeless foundation for life itself. Beyond having no temporality, love also has no specific place of residence. It isn't exclusively an internal or external phenomenon; rather it permeates presence itself. It connects us with ourselves and everything else. It is an "in between" yet ubiquitous kind of notion. The difficulty of fully exploring this emotion is that words fail to provide a grasp of the sensation capable of grounding life to awareness itself. A while ago, a friend introduced me to "Intimacy" from the journal of A. H. Almaas. As the title of the chapter suggests, here he expresses the concept of intimacy. I paraphrase: it is experienced as a simple sense of presence without a sense of self, a bare "witnessing" without a witness, perceiving without a perceiver, observing without an observer—where time doesn't exist.[10]

Once aroused, there is an early experience of love. As we expand this simple, early experience into a full range, love occurs and is experienced as joy, peace, curiosity, happiness, surprise, delight, wonder, beauty, trust, and interest. As life evolves and other aspects and emotions become part of the dance of life, love takes new forms in acceptance, empathy, awe, inspiration, relaxation, contentment, kindness, and compassion.

Experience and expressions provide a direction for emotional behavior that is active and aggressive in nature. Expressions of enthusiasm, passion, desire, affection, ecstasy, and others involve love or retain love as a guiding influence.

Returning to the baby, very early in life, interruptions occur repetitively, and patterns of recalling and remembering take hold. When comfortable patterns occur, a notion of pleasure or goodness is identified. When discomfortable patterns occur, notions of pain and responses of withdrawal become present. The discernment of patterns—and good and bad—appear very early, as experiences lead to the capacity to discriminate.

It seems like these three aspects are morphing into emotions. Are they?

Yes, they are. By adopting the basic premises of the object relations theory,[11] we see that discernment provides good and bad objects and sensations (experiences). The positive outgrowths of love are experienced when enthusiasm, passion, desire, affection, ecstasy, acts of kindness, and many others find expression. For the negative outgrowths, aggression, jealousy, spite, outrage, obsession, craving, fury, greed, hatred, rage, hysteria, lust, and many more emotions might occur. Examples of aggressive discernment are arrogance or argumentativeness.

Why emotions?

Human beings appear to be emotional beings,[12] not simply the rational beings that we might prefer to be seen as. In business, many of the difficulties that cause very costly repercussions arise from cultural attempts to rationalize almost everything.

Emotional states govern communications and one's effectiveness. Initially, many won't like this aspect of the new communications model, but such a reaction is part of any fundamental shift. When cars were introduced, there were numerous repercussions due to many people's disdain for mechanical transportation. Rightly so, as we can develop a relationship with an animal, while a machine is impersonal.

In my experience of counseling others, awareness of these three levels of existence (communication) provides a freedom and an expanded capacity for relating with themselves and others. It is this experience where I trust a contribution will be forthcoming. A benefit to the design of this new world is that the invention provides an awareness of our current emotional state each moment. This self-awareness provides for much more effective communication.

Expanded awareness

How does the invention do that?

Statistics show that face-to-face and tele-video conversations are more effective than telephone conversations because the facial muscles tell us which emotions are present.[13] We can be aware of much more when facial expressions are observed. For instance, before a baby cries, we can see it. We become aware of the emotions surfacing as his or her facial expression alters.[14] There is oftentimes a "tell" of an emotion before an awareness of the emotional shift occurs. An observant parent can interrupt an infant before he or she starts crying and can potentially shift the emotion to amusement or surprise.

Facial expressions linked to emotion have been of keen philosophical and psychological interest for over a century. Some relevant documentation appears to predate Charles Darwin's *Expression of the Emotions in Man and Animals* written in 1872.[15] Certain aspects of clinical research about the Facial Action Coding System (FACS) of Dr. Paul Ekman[16] and other psychologists might be put into action as one element of the invention as a method for identifying and informing us of our current emotional state. When we discuss this new technology, I will explain the computer expression recognition software capability. This component of the invention utilizes a camera, observing us and constantly translating our emotions into three primary states—love, aggression, and discernment.

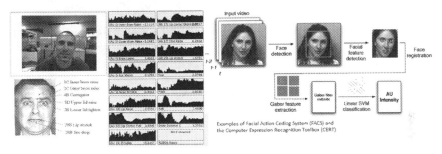

Examples of Facial Action Coding System (FACS) and the Computer Expression Recognition Toolbox (CERT)

We rely on facial expressions to a large degree. When we are aware of or attuned to what these expressions tell us, we are much more effective in communication. The invention will automatically incorporate the option of having audio/video conversations and conferences, whenever and however you desire it. Applications like Skype,[17] iChat,[18] and Google+ Hangout[19] currently use this technology. With the invention, visual interaction can be live or occur over time. Either way, the conversation is recorded, and portions can be converted to text. We are now delving directly into facets of the new technology. This will be expanded upon when we switch topics and begin part 2.

Going back to computer expression recognition, let's explore its potential to completely alter our capacity to communicate. Many of the current devices that have screens as part of their makeup also have very effective cameras that face the user. Other sensors are sometimes involved as well. Expression recognition would most likely be a gradual process, as expressions, verbal tones, word phrases, physiological changes, and the like are systematically organized over time. The specialized sensors and user-facing cameras, coupled with this application, will drastically alter our self-awareness—whether we are in communication with others or simply thinking and relating to ourselves.

Relating to ourselves sounds weird, doesn't it?

Yes, it does sound odd because, up until this invention, unless we are practicing a talk in the mirror or videotaping something for later viewing or using one of the video communication applications that we mentioned earlier, we have little experience in relating to ourselves. Except for those who have a lot of experience in this area, seeing your own face as you speak and express can be very distracting. The "expression recognition" capability provides a subtle but reliable means of letting you know what your current emotional state is. Every moment can be empowering.

Being connected to something that indicates your every change in state, just as a lie detector notes when a lie is being told, allows you to know if you are exhibiting aggressiveness, arrogance, and so forth.

Why would we care to know that?

Continually receiving feedback about when you're being too aggressive, for example, enables you to notice when you aren't connecting to the person that you are conversing with. You might be alerted when you are becoming too positional in your thinking. These obstacles in communicating reduce your effectiveness. The valuable insight and self-awareness gained from this feedback serve to combat them.

A powerful thing to note is that regardless of what method you use to converse with others or yourself, your emotions are at play. You might be speaking via video, video with a whiteboard, audio only, or text only. You are starting to get a sense of the capability of this invention, and I will expand on how this technology differs from our current understanding in part 2.

Another side of the conversation relating to the old speaker/listener model is that we are now aware of our emotional state when others express themselves. That awareness gives us further opportunity to better connect or relate to others because we are present to the interaction and our emotional reactions. If this capacity to be aware is subtle, it won't distract us but will provide a tremendous benefit.

Can you tell me about the subtle way we will become aware of our emotional state?

There are several ways, so it depends on what is most effective for you. Some want to view themselves. Each of us tailors the invention to fit our own tastes and preferences, and I will speak more to that in the next section, part 2. For the moment, let's look at a couple of ways of providing feedback. One way is the visual. Emotionally, we have a blend of three primary states:

- ✓ Peaceful, connected, joyful, caring

- ✓ Expressive, aggressive, poignant

- ✓ Pensive, thoughtful, withdrawn, explanatory

One possible way of illustrating these different primary emotional states is to utilize the primary colors: blue, red, and yellow. As there are various blends of the emotional states, we can also utilize the secondary colors (and numerous blends of the primary colors) as a means for feedback. Variations of purple, orange, and green might come into play, as changes in emotional state are represented by changes in the shade of color.

Where would these indicators be?

An indicator could appear under your picture, or it could color the border or background of the screen itself. Each of us gets to choose the best indicator placement. Another feedback aspect will denote whether we experience like or dislike and whether we find our emotional state pleasant or not. These can be indicated by an alteration of the screen's brightness. In addition to the visual indicators, an audio (and/or haptic) component might also be employed. The nature of our interface using this technological adaptation for communication is almost entirely up to us. Some of us like music. In another design, background music shifts alongside the changes in emotion. There is a wealth of options in designing our new environment.

I am having a hard time grasping this invention. When can we discuss the technology?

We'll be discussing the technology in part 2, "A New Way to Relate to Technology," chapter 4. Part 2 will also include quite a bit more about communication, as well as the reconstruction of other ancient paradigms.

1. Humberto R. Maturana and Francisco Varela, *The Tree of Knowledge* (Boston: Shambhala Publications, 1987).

2. Albert Einstein, *Physics and Reality*, trans. Jean Piccard (n.p.: Prof. M. Kostic, 1936), 349–382, http://www.kostic.niu.edu/Physics_ and_Reality-Albert_Einstein.pdf.

3. Ludwig Wittgenstein, "A Lecture on Ethics," (1929), accessed May 22. 2012, http://www.galilean-library.org/manuscript. php?postid=43866.

4. Martin Heidegger, *Letter on "Humanism"*, transl. Frank A. Capuzzi, (1949), 239, http://www.archive.org/details/ HeideggerLetterOnhumanism1949.

5. Michael R. Gilmore, "Einstein's God: Just What Did Einstein Believe About God?," *Skeptic Magazine*, vol. 5, no. 2, (1997), 62, http://www.skeptically.org/thinkersonreligion/id8.html.

6. Daniel Gilbert, *Stumbling on Happiness* (New York: Knopf, 2006).

7. François-Marie Arouet de Voltaire (1694–1778).

8. Alexandra Tyng, *Beginnings: Louis I. Kahn's Philosophy of Architecture* (New York: John Wiley & Sons, 1984), 177.

9. Humberto R. Maturana and Francisco Varela, *The Tree of Knowledge* (Boston: Shambhala Publications, 1987).

10. A. H. Almaas, *Luminous Night's Journey: An Autobiographical Fragment* (Boston: Shambhala Publications, 1995).

11. Melanie Klein, *The Psychoanalysis of Children*, trans. Alix Strachey (New York: Grove Press, 1960), 13.

12. Humberto R. Maturana, *Metadesign.* (August, 1, 1997), accessed May 22, 2012, http://www.inteco.cl/articulos/006/texto_ing.htm.

13. "Mehrabian's Communication Research," accessed May 22, 2012, http://www.businessballs.com/mehrabiancommunications.htm.

14. Malcolm Gladwell, *Blink: The Power of Thinking Without Thinking* (New York: Little, Brown, 2005), 197–214.

15. Charles Darwin, *The Expression of the Emotions in Man and Animals* (Stilwell, KS: Digireads.com Publishing, 2005).

16. "Dr. Paul Ekman." Paul Ekman Group LLC., accessed May 22, 2012, http://www.paulekman.com/.

17. "Skype," Skype, accessed May 22, 2012, http://www.skype.com/intl/en-us/home.

18. "All Applications and Utilities," Apple, accessed May 22, 2012, http://www.apple.com/macosx/apps/all.html#ichat.

19. "Google+ Hangouts," Google, accessed May 22, 2012, http://www.google.com/+/learnmore/hangouts/.

CHAPTER 2
AN ONTOLOGICAL VIEW
OF COMMUNICATION*

The linear communication discourse

Where does our current communication paradigm come from? How did we inherit linear communication?

The first suggestion of linear communication and a linear view of reality appeared six hundred generations ago in Africa. Village life began to form. Social interactions began developing a complexity of recurring patterns. This complexity, for the most part, continues today with little change.

Historically, there are many types of cultures: warlike, cooperative, competitive, matriarchal, and patriarchal. These developed over time out of the reoccurring interactions that village life required, and many such developments still exist today.

These early societal flavors were made possible by basic patterns, causing an evolution from familial and coven to village and tribal formations. The domestication of plants and animals marked one of the first shifts in technological space in most areas. New political and philosophical patterns arose from the simple foundations of early village life. There appear to have been some common structural patterns.

Underneath all of these nuances, something else related to the complexity of the abundant social interactions required by community life shifted. In the main traditions of both the East and West, if dissimilar foundations existed, they have been lost or have very little influence today.

The most basic reconstruction of these first primitive happenings came with the consideration of building a close family or coven environment. Little of this environment was expressed in words. A look or gesture was required. It was this basic familiarity as a means of coordinating that got lost in a village environment. Village life thrust a vast complexity that required a new way to coordinate. This expansion into expressing words was fairly consistent throughout the world.

There was an ease to keeping track of a small coven or family and their possessions. In a village, interaction expanded in a way that was consistent even in the different regions. Counting—a linear means of bringing order— came into existence.

There were simple roles to play in a family or coven. In tribes or villages, a new structure for specialization joined the scene. The use of specialization in communal life is pretty consistent worldwide. What took place eons ago gives us the basic foundation for our reality and the previously uncontested foundation for linear communication and accounting.

Our inheritance persists to this day. There is a correlation between our inherited understanding of and structures for communication (speaker → something deliverable → listener) and the method of communication that we practice today. The linear path of communication and activities involved for its coordination has created a very chaotic atmosphere for the completion of projects, events, and objectives. The rapid expansion of specialization in the last few decades adds to this chaos.

With communication via chat, e-mail, telephone, and so on, it seems that linear communication is the accumulation of conversations. Linear processes seem to be almost exclusively the method used for communication and the process orientation that something passes through. Changes come from the in/out motion of process orientation.

From linear communication come linear management, organization, and process orientation. Financial and management accounting also stem from this area (e.g., customer relations management, supply chain management, and project management). We will go into linear processes more deeply in part 3. Beyond linearity, more to the point is how inadequate our current communication appears when we develop projects, theories, and cooperatives. These endeavors are fraught with breakdowns.

Communication limitations

Can you illustrate where the current communication paradigm breaks down?

To illustrate, consider the two following examples:

> Man is spirit. But what is spirit? Spirit is the self. But what is the self? The self is a relation which relates itself to its own self, or it is that in the relation [which accounts for it] that the relation relates itself to its own self; the self is not the relation but [consists in the fact] that the relation relates itself to its own self. Man is a synthesis of the infinite and the finite, of the temporal and the eternal, of freedom and necessity, in short it is a synthesis. A synthesis is a relation between two factors. So regarded man is still not yet a self.
> —Sören Kierkegaard, *Sickness Unto Death*[1]

> Note: The brackets above occur in the book.

> Everything said is said by an observer to another observer that could be him or herself.
> —Humberto R. Maturana, "Ontology of Observing"[2]

These two quotes can easily represent a multitude of brilliant philosophers, scientists, and other rigorous experts. It is the speaker/listener paradigm that has brilliant thinking and quotations appear as a struggling articulation. Aristotle captured the edges of communication thousands of years ago.

Until we expand communication beyond this single-dimensional thinking, little new can be expressed. Explanations of how life occurs can appear as self-referential. However, from three dimensions, that isn't the case. When communication occurs in three dimensions, spirit, self, and observer aren't defined. They simply occur in a dimension, one that is timeless and formless. It points to relating (a timeless, formless quality)—there is no location or comprehension associated with relating and eternity.

If we wish to discover the nature of human existence, three dimensions (levels) convey awareness:

✓ eternal

✓ now

- ✓ temporal

  or

- ✓ relating

- ✓ experience

- ✓ cognition

From three dimensions, you can read the Kierkegaard and Maturana quotes, view the awkwardness, and experience its expression in a new way!

When viewing life as emotional entities, it can be useful to view those emotions that occur as the most real. The timeless and formless aspect that we can point to as relating appears to be the vast foundation underlying existence. Consider your question in the last chapter: "This current view of communication seems real. Why mess with it?" What we have distinguished as discernment deserves a deeper visit into reality.

What do you mean by "a deeper visit into reality"?

From the communication model, reality is discernment. I desire taking a nonorthodox approach. Normally, ontology deals with objective reality. There are already many arguments ontologically.

- ✓ Thoughts give temporality—a past, present, and future.

- ✓ Temporality gives us motion, and a world that is real comes into our understanding.

- ✓ Our lives occur as a continuing spectacle, and the perception of change appears real.

- ✓ Language and discourses—the history of humanity gives us culture, habits, and an automatic ease of functioning each moment.

This aspect of life, this emotional level, provides what is important to us. That the world seems "real" gives us a parallel similar to the movie *The Matrix*. We are deluded. To effectively go down the rabbit hole, we will take a trip in the world of physics, a discourse most of us are somewhat familiar with. Einstein said, "People like us, who believe in physics, know that the distinction between past, present, and future is only a stubbornly persistent illusion."[3]

Albert Einstein—the physics of it

What are we going to explore—physics and reality?

Yes, we are. I am interested in leaving philosophy for just a moment and seeing where we end up. I will be using Albert Einstein's discourse to direct the way. Here is the first premise, as science historically usually discounts philosophy.

> Such might indeed be the right thing at a time when the physicist believes he has at his disposal a rigid system of fundamental concepts and fundamental laws which are so well established that waves of doubt cannot reach them; but, it cannot be right at a time when the very foundations of physics itself have become problematic as they are now. At a time like the present, when experience forces us to seek a newer and more solid foundation, the physicist cannot simply surrender to the philosopher the critical contemplation of the theoretical foundations; for, he himself knows best, and feels more surely where the shoe pinches. In looking for a new foundation, he must try to make clear in his own mind just how far the concepts which he uses are justified, and are necessities.
>
> The whole of science is nothing more than a refinement of everyday thinking. It is for this reason that the critical thinking of the physicist cannot possibly be restricted to the examination of the concepts of his own specific field. He cannot proceed without considering critically a much more difficult problem, the problem of analyzing the nature of everyday thinking.
>
> On the stage of our subconscious mind appear in colorful succession sense experiences, memory pictures of them, representations and feelings. In contrast to psychology, physics treats directly only of sense experiences and of the "understanding" of their connection. But even the concept of the "real external world" of everyday thinking rests exclusively on sense impressions.

I want to stop here for just a moment and check in with what you are seeing. If you recognize that Einstein wrote this over seventy-five years ago (in 1936), it will give you a perspective of his brilliance. However, all of us inherited the same discourse of communication, and it is only today that we will begin to break free from this old "real" paradigm. What do you see about what Einstein is deliberating?

He seems to be bringing the knowledge of physics into question. When he speaks of "experiences" and "understanding," it sounds like you. He also directs his thoughts on the "real external world" as a concept. Is he saying this?

As much as I have studied Einstein, as I was preparing and verifying the credibility of the quotes I was using for this book, I came across this work, *Physics and Reality*, for the first time. I don't think he sounds like me. It is I who has studied his work. I think that exploring his work throughout my life and examining his methodology of thinking has played a big part in the freedom and courage for me to dismiss the foundations of our existence and rewrite and construct a new foundation.

> Now we must first remark that the differentiation between sense impressions and representations is not possible; or, at least it is not possible with absolute certainty. With the discussion of this problem, which affects also the notion of reality, we will not concern ourselves but we shall take the existence of sense experiences as given, that is to say as psychic experiences of special kind.

It is here, Einstein begins speaking of "awareness." That is something that we will address shortly as he will utilize the term "comprehensibility" as he labors in defining awareness.

> I believe that the first step in the setting of a "real external world" is the formation of the concept of bodily objects and of bodily objects of various kinds. Out of the multitude of our sense experiences we take, mentally and arbitrarily, certain repeatedly occurring complexes of sense impression (partly in conjunction with sense impressions which are interpreted as signs for sense experiences of others), and we attribute to them a meaning—the meaning of the bodily object. Considered

logically this concept is not identical with the totality of sense impressions referred to; but it is an arbitrary creation of the human (or animal) mind. On the other hand, the concept owes its meaning and its justification exclusively to the totality of the sense impressions which we associate with it.

What are you hearing now?

Einstein has basically said that our thinking gives meaning, but this interpretation is not really what we are sensing.

The second step is to be found in the fact that, in our thinking (which determines our expectation), we attribute to this concept of the bodily object a significance, which is to a high degree independent of the sense impression which originally gives rise to it. This is what we mean when we attribute to the bodily object "a real existence." The justification of such a setting rests exclusively on that fact that, by means of such concepts and mental relations between them, we are able to orient ourselves in the labyrinth of sense impressions.

Now I think he is saying that we are predisposed to what we sense.

These notions and relations, although free statements of our thoughts, appear to us as stronger and more unalterable than the individual sense experience itself, the character of which as anything other than the result of an illusion or hallucination is never completely guaranteed. On the other hand, these concepts and relations, and indeed the setting of real objects and, generally speaking, the existence of "the real world," have justification only in so far as they are connected with sense impressions between which they form a mental connection.

The very fact that the totality of our sense experiences is such that by means of thinking (operations with concepts, and the creation and use of definite functional relations between them, and the coordination of sense experiences to these concepts) it can be put in order, this

31

fact is one which leaves us in awe, but which we shall never understand. One may say "the eternal mystery of the world is its comprehensibility." It is one of the great realisations of Immanuel Kant that the setting up of a real external world would be senseless without this comprehensibility.

In speaking here concerning "comprehensibility," the expression is used in its most modest sense. It implies: the production of some sort of order among sense impressions, this order being produced by the creation of general concepts, relations between these concepts, and by relations between the concepts and sense experience, these relations being determined in any possible manner. It is in this sense that the world of our sense experiences is comprehensible. The fact that it is comprehensible is a miracle.

In my opinion, nothing can be said concerning the manner in which the concepts are to be made and connected, and how we are to coordinate them to the experiences. In guiding us in the creation of such an order of sense experiences, success in the result is alone the determining factor.[4]

I am not surprised that my model not only is validated in Einstein's discourse but that it opens up the artistic quality of life. When speaking of this new communication model, I often use an orchestra as an analogy. (I appreciate this correlation, as the conductor's perspective provides order and structure to art.) At any time, the brass, winds, and strings provide an awareness of music and its beauty. The three levels of communication provide intuitiveness and awareness. Yes, that we learn to coordinate these three levels with some effectiveness in the first two or three years of life is a miracle.

Awareness

What do you mean when you use the term "awareness"?

Awareness is mostly stipulated as a cognitive function, which is aligned with its dictionary definition of cognition and knowing. The term also has spiritual and philosophical roots. From the spiritual aspect of consciousness, a very diverse notion of awareness has come to the foreground. Is awareness cognitive or experiential or even eternal?

This notion of awareness is a quandary that we are beginning to unravel. Awareness, from a view of comprehension, seems to have shown up as an expansion of a cognitive function. While we are connected and exist in a relational space, and while we continually experience the world that our environment contributes, the tiny window of experience that we are aware of and the fabric of reality in which our awareness occurs have some sort of epistemological characteristics.

Experience and cognition are different domains, different planes of being in the world. Cognition, knowledge, and language are a portion of awareness, but the richness of life isn't found on this level. Rather, this level gives us what is important.

When a curious experience and cognition touch each other in an internal phenomenon, the blending of these two levels gives a sense of harmony—awareness. Awareness, thinking, cognition, and discernment have congruency with each other. Harmony, love, and awareness (alertness) are also congruent.

There are also multitudes of moments of awareness where we act without thought, yet we are aware. There are those moments we get in touch with stillness or art, beauty, and expression. The intention of this invention is to expand awareness, giving us access to these moments and a compass to appreciation and effectiveness in life. This is how awareness is utilized, without a refined definition, rather as a blend or an orchestration of the three levels.

Does awareness undermine rationalization?

Dr. Fernando Flores[5] and Dr. Humberto Maturana[6] distinguished observers, contexts, and cognition. When viewing their research and conclusions through our existing paradigm of communication, a new paradigm (communication relativity) opened. This opening permitted and empowered a simple proposition—one that disclosed and revealed what provides observers, contexts, understanding, and cognition.

I saw cognition as responsible for the limits we find ourselves in. This new world gives you an opportunity to start thinking, communicating, and functioning from a simpler starting place—a place where the world we live in and the structure for relating converge.

Cognition provides the width and depth of reality. At any given moment, our slice of awareness appears small, giving us the illusion that reality must be vast and expansive. Our memory of the past and anticipation of the future, observation of current developments, and our awareness of movement combine to give us the sense of a complete environment—reality.

Albert Einstein leads us to ponder what we know and remember in saying, "Every reminiscence is colored by today's being what it is, and therefore by a deceptive point of view."[7] He also rationalized, very succinctly, where cognition originates.

> Many people think that the progress of the human race is based on experiences of an empirical, critical nature, but I say that true knowledge is to be had only through a philosophy of deduction. For it is intuition [and awareness] that improves the world, not just following a trodden path of thought. Intuition makes us look at unrelated facts and then think about them until they can all be brought under one law. To look for related facts means holding onto what one has instead of searching for new facts. Intuition is the father of new knowledge, while empiricism is nothing but an accumulation of old knowledge. Intuition, not intellect, is the "open sesame" of yourself.[8]

Einstein spoke of those moments when his breakthroughs and theoretic models appeared just as he imagined and wondered about what he experienced.

It was blending the teachings of Dr. Fernando Flores and Brian Regnier that produced a very intriguing awareness of cognition as a historical discourse. I gained an experience of discourse that transformed my vast knowledge of discourses. From this perspective, considering cognition, language, and the world, the constriction and limits of cognition and reality occur as experience.

It was my fascination with exploration and the world of relating[9] that altered my awareness, intuitiveness, and imagination. It was here where mankind's issues, on both a micro and macro level, were revealed. Our struggles to understand, to control, and to seek the elusive gained clarity.

Out of this realization, a model depicting life's relationship with the world and the simplicity of observation and effectiveness was born. This model is communication relativity.

Cognition dictates our judgment of what is important, but what we are aware of without assistance is very limited. It is a tiny sliver of our existence. Communication relativity is built on the evidence that we exist in three distinct planes simultaneously, and being aware of these three planes simplifies and expands our capacities.

By gaining an expanded awareness, we can more easily throw away the unexplored assumptions under which we understand, act, and experience life. This new technological space will give us newfound freedom as we are transported into an environment of clarity and structure with a sense of being at home with ourselves, others, and our circumstances.

1.   Sören Kierkegaard, *Sickness Unto Death,* vol. 19 (Radford: A & D Publications, 2008).

2.   "Ontology of Observing: The Biological Foundations of Self Consciousness and The Physical Domain of Existence," Humberto Maturana, Instituto de Terapia Cognitiva INTECO–Santiago, accessed May 22, 2012, http://www.inteco.cl/articulos/004/texto_ing.htm.

3.   Tim Folger, "Newsflash: Time May Not Exist," *Discover Magazine,* June 12, 2007, http://discovermagazine.com/2007/jun/in-no-time/article_view?b_start:int=1.

4.   Albert Einstein, *Physics and Reality,* trans. Jean Piccard (n.p.: Prof. M. Kostic, 1936), 349–382, http://www.kostic.niu.edu/Physics_and_Reality-Albert_Einstein.pdf.

5.   Terry Winograd and Fernando Flores, *Understanding Computers and Cognition: A New Foundation for Design* (Norwood, NJ: Ablex Publishing, 1986).

6.   Humberto R. Maturana and Francisco J. Varela, *The Tree of Knowledge: The Biological Roots of Understanding* (Boston: Shambhala Publications, 1987).

7.   American Association of Physics Teachers, Gerald James Holton, American Association of Physics Teachers. Committee on Resource Letters, American Institute of Physics, *Special relativity theory: selected reprints* Ann Arbor: Published for the American Association of Physics Teachers by the American Institute of Physics, 1963, 14.

8. William Hermanns, *Einstein and the Poet: In Search of the Cosmic Man* (Brookline Village, MA: Branden Press, 1983), 16.

9. "Partnership Explorations Course," Landmark Education Wisdom Division, accessed May 22, 2012, http://www.landmarkeducation. com/Graduate_Center/About_Graduate_Programs/The_Wisdom_ Courses/Partnership_Explorations_Course.aspx.

Chapter 3
Communication, Under the Hood*

Communication relativity and awareness

Can you say more about communication relativity?

Communication relativity gives each of us feedback about our current communicative temperament. How our current state is detected and distinguished will be discussed in terms of "interactivity awareness" and "interactivity awareness support." The feedback shares similarities with cybernetics and agent feedback loops of intelligence, but it is simplified and based on a new communications model. This new model stipulates that there are three levels of communication.

In having this discussion, we are discussing the technology possibly more deeply than the introduction allowed for. What will be expanded is the notation in the invention diagram in both the introduction and chapter 1—"Real-time Feedback of Effectiveness: A Novel Communication Breakthrough." As discussed, these levels are summarized as the three major emotional states: love, aggression, and discernment. The blending of these three levels of communication would enhance effective communication. The visualization inferred is to have love be blue, aggression red, and discernment yellow; also, bad is dim, and good is bright. Here are a few examples of the blending of these levels and use of color as a visual aid:

- ✓ The background is bright and has a blue hue when speculating.

- ✓ A very alert and aware state would be bright with no distinct hue.

✓   When making plans or requests, the background is neutral or slightly red, indicating either a balanced or a mildly aggressive state.

✓   When dealing with problems, a bright background would be constructive—signifying that you are looking for an opportunity. A light green background could be related to thinking about ideas or possible solutions.

If something unfavorable occurs, the screen's hue grows dull. Over time there is an opportunity to learn to express this dislike in such a way that green rather than orange (arrogance) is present. This feedback gives us access to the three different levels of communication.

What are interactivity awareness and awareness support?

Interactivity awareness is made available by a camera that faces you. It provides an emotional feedback and requires devices capable of supporting communications. The desktop provides interactive elements and a viewing screen to illustrate the feedback. Part of the awareness has to do with the content or context of the desktop environment and the interaction and interactivities occurring.

Interactivity-awareness support analyzes your emotional state. This support learns what your facial expressions are.

Beyond facial recognition, many other bodily functions can indicate an emotional awareness. There is no conceivable way to capture all the methods that could support and enhance interactivity awareness. The task for entity relativity is to learn what is relative for each of us. Voice inflections, words, and phrases are just a few things captured. Recognizable patterns are transferred to interactivity awareness support.

Distinguishing each emotional state, the terminology, interests, and so forth develops a type of profile for you. A historical capacity to anticipate an awareness of activities and preferences gives the virtual environment the ability to increasingly mirror virtual reality to your physical reality. The bank of knowledge becomes more precise over time as awareness of your bodily changes, inflections, and actions is continually refined.[1]

Three distinct communication levels

You are talking about emotional states. Can you say more about what is measured and the different levels of communication?

In describing the three different aspects of communication, it's important to express that they are completely dissimilar in nature.

One level is so subtle that it does not exist in the physical world. Sciences are unable to capture its essence. This level is elusive—like magnetism and its effect on ferrous metals—without discernible evidence aside from its impact. Also similar to magnetism, it is permanent and unaffected in space and time. It remains connected, regardless of orientation. Orientation can still serve to characterize a connection though. The first level is timeless and formless and can never be observed. It is more like a relational space from which observation can occur. There are facial and bodily changes that designate the occurrence of this level, just like iron filings can illustrate the presence of a magnetic field.

The next level is poignant. It is expressive and experiential, with qualities like a spark: a momentary happening that has internal and/or external qualities. It is a phenomenon that happens in the moment as it touches the other two levels. The second level includes much more than just our senses. There are specific expressions and bodily changes that indicate that this level is present.

The third level compiles the moments of the second level and creates the cognizance of movement. This level recognizes many kinds of patterns, but this recognition/cognition is strictly an internal phenomenon. Pattern recognition leads to more sophisticated levels of cognition (learning, comprehension, and memory). This level is a temporal discernment, with a sense of past, present, and future. It brings forth reality, discourse, and cognizance. This level is where our plans, thought processes, and what feels important appear. This level is also physically illustrated.

The third level is influenced by the second. In fact, it has some dependence on the previous level. Our awareness of a tiny fraction of the second level, coupled with the act of thinking about our thinking (the third level), gives us our reality, our existence, our life.

We never fully comprehend our moments on the second level, but we can remember them, put them together, and pull meaning or recognition from them. We can't experience expressions of the first level, but this level and its expression influence our experiences. Afterward, we often remember these influenced experiences. Art, poetry, music, and other forms of expression are influenced by all three levels—this illustrates communication, and consequently the orchestration of the three distinct levels of communication.

How does each device remember these different levels?

They don't. The new environment, the virtual companion, utilizes the devices as a conduit and a means to connect with a virtual environment but has no need to remember anything. We are getting into parts 2 and 4. Let's move on to the next section.

1.   This profile is described as the entity provider in the notes after chapter 10.

Part 2
A New Way to Relate
to Technology

CHAPTER 4
WHAT IS TECHNOLOGY, REALLY?

An intuitive relationship

What do you mean by a new way to "relate" to technology?

I mean exactly that! From the time of the caves (six hundred generations ago) onward, we have maintained a consistent relationship with technology. Our tools, instruments, and weapons have become more sophisticated, and we have developed all kinds of marvelous vehicles. We've invented wheels and harnessed fire. Air flight, space travel, electricity, and electronics have provided many modern inventions. Despite changes, these still exist as tools, utensils, and weapons to be used.

A rare exception is seen in the implants and artificial organs that replace living organs. The introduction of further devices and robots expands the capacity to replicate organs and human functionality. There is one aspect of technology, however, that we rarely think of as technology. Most animals, insects, and other life forms exist in a technological space. This technological space is their dwelling space. Human beings have capitalized on using tools, but what would be available if our tools resided where we dwell?

What do you mean?

An illustration of this, which few know about, relates to the intuitive awareness that fighter pilots have in their cockpits. Unless you have had that specific experience, think about your dwelling place. We have specific instruments, furnishings, tools, and devices for the different rooms of our homes. We also have controls specific to each device. Most things stand different, separate. Most applications operate separately, though sometimes we have found it useful to merge certain aspects of activities. Technology is oriented in two different ways: as objects and applications or as dwellings.

In life, however, we design instruments with specific applications, and everything lives as a tool or instrument to serve its own separate purpose. Our attention shifts from device to device, system to system, throughout the day, and life becomes extremely device-centric. Consequently, we are distracted by numerous processes that are not even necessary parts of living life. These processes are additional, superfluous. There is a divide. We repetitively interrupt our lives to record data so that we have access to information in the future.

I really dislike having to enter information and account for things. Can you give me a simple example that will give me a sense of what you are talking about?

The kitchens of our dwellings provide a certain environment for our family. This environment empowers us to take care of our family. I am not a cook, but my wife is. When she cooks in the kitchen, coordination takes place. An awareness of time, temperature, and other subtleties consistently converge into a wonderful meal. In the kitchen, a dance takes place; utensils, stovetops, ovens, and food are at play. At times, I walk into this symphonic creation. The creativity has a magical quality to it. In this description, we see a simple example of how space can be designed for its relative activities and purposes. The design of a space can support an area of life that is important to the family.

The rooms of our dwellings provide environments in which to coordinate our activities and interactions for living life. Rooms are also designed to fit different people. Our dwellings are not device- or application-centric. They are designed to provide for specific interests and concerns.

How does my dwelling relate to the rest of my life?

The more we look at what is important to us, the more we can understand how these various rooms get organized. There are areas of our lives that are important, that we are interested in, and, at times, concerned about. In each room, the present objects don't translate readily into the makeup of other important areas. Even the means of maintaining various spaces are different, as we now have bathroom cleaners, kitchen cleaners, carpet cleaners, and hard floor care.

If you leave the world of device- and application-centricity and enter a world that centers on what is important or of interest, a whole new environment opens up.

I don't understand. How can technology exist without devices?

The shift is in orientation, not the use or disuse of applications and devices. Certain devices are called for, depending on where we physically are. There are different aids and instruments for every important area of life. A number of people and resources are also vital to these different fields of life. But what if the devices were nondiscriminating? What if, no matter where we are—at home, at work, in the car, at a restaurant—the devices are designed to gives us a seamless way of relating through *a virtual world that mirrors* what is important to us each moment? Devices, then, are no longer something to use or access; they provide an environment that constantly supports our current interests, conversations, and activities.

One noticeable thing that spans all areas of importance is scheduling, the means of maintaining numerous interests effectively. Scheduling records specific, important events, projects, and plans and allows us to coordinate among various involvements in our lives. The act of coordination is a critical part of life. Soon, we will delve into scheduling as an element of structure and the new method of accounting in an environment where processes give way to structural design. For now, I'll speak specifically about a new technological space and the important areas of life.

Attention on what is important

What do you mean by important areas of life?

This technology brings what is important to the foreground. What is important to us often appears in the form of interests and concerns and can be seen in what we account for, keep track of, and measure. These interests and concerns enter the stage as we interact with both others and ourselves. We speculate about, plan, schedule, and commit time to the things we desire. We discuss these things. We measure them. We notice changes and revisions to them over time. Within this technological space, we deal with, communicate, and account for what is significant and tangible.

People coordinate and collaborate through interactions. In the last few dozen years, interactions have been recognized as the vehicle for progress. What we have inherited as accounting has lagged behind in this awareness. We deal with accounting in terms of transactions. With the tremendous speed made possible by computerization, the time between interactions and transactions has diminished tremendously in most cases. Regardless of the almost simultaneous occurrences, however, no transaction occurs without interactions. In part 3, a new world of accounting will be discussed specifically and more fully.

What distinguishes different areas of life?

There are two common divisions of interests and concerns.[1] One is oriented toward our relationship with ourselves and the world. The other is oriented toward our interactions in organizational, communal, or societal aspects of living life. The first pertains to the interests and concerns of each individual, relative to you and your experience of life. The second is based on the interests and concerns of organizations or any group that includes more than one individual. Some basic organizations are families, neighborhoods, businesses, and governments.

Chip Wilson, founder of lululemon athletica, a very successful apparel organization, said something to the effect that new employees are integrated into an orientation where they become "unmuddled." He attributes effectiveness and extremely low turnover to this business philosophy and education. What he illustrated looked very similar to many of the common interests and concerns that are important to us. When his employees' lives gain clarity, it translates into business interest effectiveness.[2]

What are some different interests and concerns? Can you be specific?

There are many common interests and concerns for humanity. Here are several of these:

- ✓ Family

- ✓ Friends

- ✓ Money

- ✓ Work

- ✓ Health

- ✓ Education

- ✓ Spirituality

- ✓ Sexuality

- ✓ Respect and dignity

- ✓ Wealth and property

- ✓ Careers and contributions

- ✓ Play, fun, art, recreation, and hobbies

- ✓ Circumstances, current issues, and plans

- ✓ Causes, politics, belonging, and communities

While this list includes many important sectors, interests and concerns aren't specifically limited to those listed.[3] These categories might very well have subsections and are further distinguished by projects, events, and meetings. In these interests and subinterests, people and resources become available, and tools to support the specific areas are readily accessible.

Organizational interests

Both individuals and organizations occur as identities. A connection to the interests and concerns of communal entities is created by individuals and passed to their organizations. A work entity might house the structure of interests and concerns for a business. Family might house the interests and concerns for the household. For instance, specific interests of a family may include meals, homework, activities, family time, couple time, household responsibilities, and accountabilities.

A way to demonstrate the shift from application- to interest-centricity is the current practice of browsing the Internet. When searching for specific interests, questions, and inquiries, these activities are supported by search engines, social sites, and Wikipedia. In this new environment, the process of going to a search site and searching the Internet disappears. Instead, resources for a specific interest are intuitively offered. Other examples of processes that no longer exist are e-mail, getting directions, and locating establishments. In summary, we will no longer exist in a device- and application-centric environment.

We will be living in an interest-centric world innately in tune to what is important. Within this world, we get to design our environment within each interest reflecting our way of viewing things. The only influence that could introduce structures or limit our autonomy comes from those areas or interests that are specifically organizationally derived.

What do you mean by our being limited? How will our autonomy be compromised?

Up until this new technology, organizations have spent tremendous amounts of money training new people. The learning curve and other difficulties for employees put a drag on organizational growth. Within the category of work, a ready example is your working for a business. With this job come certain interests, concerns, and designs for the organization, which will exist in the background.

> We define organic order as the kind of order that is achieved when there is a perfect balance between the needs of the parts, and the needs of the whole.
> —Christopher Alexander[4]

As you coordinate in the work environment, the interests of the business gain certain influence that might intrude upon or seem to diminish privacy in the organizational areas of work. In working, your association with the organization could bring areas of interest and concerns like the following:

✓ Sales

✓ Operations

✓ Finances

✓ Marketing

✓ Legal matters

✓ Technology and facilities

✓ Research and development

✓ Management

This list doesn't represent any limits to the different domains that are important for businesses. The structures of communal entities like families, churches, and governments might differ completely from that represented by work.

What are some of the benefits of this technology for organizations?

This question leads us into the third section of this book—accounting. Currently, there is a certain degree of documentation and use of information found in manuals, explanations of processes, rules, regulations, and cultural design. Acclimation to these complexities is costly, as it requires the time and effort of employees. Training and indoctrination must be designed, planned, scheduled, and taught. This will shift as businesses' rules and accountability begin to occur as design parameters and not as information or something tangible.

This new technological world brings culture, policies, processes, organization, and more into an interactive realm where research, design, and structure occur amid the resources. Supportive elements dance between the database, design, and active systems with cloud-based and other interactive collaborations. Chapter 7 will bring an awareness to infrastructure and processes that become superfluous in accounting, measuring, and tracking. A paradigm shift alters the structure, simplifying organization in this new world.

1. Guillermo Wechsler,"Ontological Design," Service Design gw (blog), February 15, 2007, http://servicedesigngw.blogspot.com/2007/02/ontological-design.html.

2. "Conference for Global Transformation," accessed May 22, 2012, http://wisdomcgt.ning.com.

3. C. F. Flores,and M. Graves, "Domains of Permanent Human Concerns," (unpublished manuscript, Berkley: Logonet Inc, 1986), e-mailed text document.

4. Christopher Alexander, Murray Silverstein, Shlomo Angel, Sara Ishikawa, and Denny Abrams, *The Oregon Experiment* (New York: Oxford University Press, 1975), 14.

CHAPTER 5
TECHNOLOGY, ANTHROPOLOGY, AND PHILOSOPHY*

The essence of technology as a discourse

You use the term "technology" differently?

Yes, I do. Martin Heidegger and many others spoke of technology in terms of our inherited paradigm. Heidegger spoke of it as relating to the "essence of technology," which is distinct from technology itself. He referred to our relationship with technology as both a means and utility. From this perspective, technology occurs as instruments and anthropological utilities. It occurs as instruments or apparatuses that require repetitive use in order to gain mastery. He does, interestingly enough, open the door for technology to be considered as art and poetry.[1] But, he stopped short. In reconstructing our relationship, we can distinguish a new technological paradigm—technological space.

What does technological space mean?

A technological space is simply our relationship with technology. Technological spaces might also refer to historical advancements, where humanity has gone through different spaces or eras of technology. I briefly discussed this in the beginning of chapter 4, "What Is Technology, Really?" Our ancestor's first technological space six hundred generations ago arose from leaving caves and forming social structures of village life. Language was enhanced technologically as social interactions became more complex.

Our technological involvement moved from the use of knives, spears, pottery, and harnessing of fire to building structures and domesticating plants and animals. Simple tools used two million years ago changed little until the domestic revolution, around 10,000 BCE. As villages formed, trade within the community began taking form.

When larger settlements developed around 6,000 BCE, metal and clay utensils were manufactured. Objects took on geometric designs. Bricks allowed sustainable structures to be built. Minerals of the earth were utilized. In China, even coal was refined. A degree of sophistication in tools and instruments was needed to deal with a more urban lifestyle. Very rudimentary forms of accounting occurred.

Less than 3,000 years later, the technological space and the language of technology quickly expanded. Something happened when stories were written and history was recorded. It changed the way humans thought and how they communicated, remembered, and preserved their thoughts. Scientific, religious, and philosophical thoughts were recorded and defined and gained permanence. Trade took hold. Ships were built as commerce took to the seas, and further technological advancements were recorded. The first written artifacts indicate the use of accounting and give evidence that the exchange of coins and money appeared a little later in history.

By 1,000 BCE, warfare was prevalent, and social organization took the form of empires and kingdoms. When cultural prominence asserted itself, metallurgy gained sophistication; iron was developed. Significant advances were also made in the arts and sciences. Thinkers like Aristotle and Plato marked this moment in history. Besides significant philosophic advances, major religious leaders and movements came into existence. The founders of Christianity, Buddhism, Confucianism, and Islam were born.

A technological space of "representative meaning" took hold next. During this vast period of time, everything was seen to relate to God and spirits, good and evil. The world was problematic, so the church provided direction. Godliness and religious leaders were revered.

As explorers chipped away at old ideas, a new thinking began to emerge, and the masses held these truths closely. New ideas from religion and science, especially the concept of the world as a globe, met tremendous resistance. There came a divide between those who ventured out to colonize new lands—to express and practice new beliefs—and those who firmly held on to the status quo.

Starting around 1650 CE,[2] a cognitive technological space for classifying new things and ideas occurred. That which resembled what was "already known" became spurned. The importance of analysis was elevated. This world provided an order and a measure to everything. Everything was examined, deciphered, and then identified as specific elements.

This period where everything was categorized provided for the birth of encyclopedias. A compulsion for structure gave a specific order to life. Language itself became a medium for transmitting the truth as life came into view in a clear and precise manner. Practicality, evidence, and certainty were valued in this new technological space.

Then, after 1800 CE, a technological space for systems was born. An expansive period began where bigger and better opportunities were consistently sought after. Industrialization and greater physical technology abounded. Architectural creativity and education became central concerns. Relating to language as merely representational was too limiting. Career specialties and specific education came into existence as standard practice. Complexity, improvement, and material things were valued.

Science could no longer be limited to describing life, the universe, or other things as simply materialistic ideas. Instead, humanity gained new understandings of things based on their systems and function. We have entered into the rationale of knowing things by the space they inhabit, by the structure that requires them to continue.

Are you suggesting we are entering into a new Foucault episteme?

As we explore this modern realm, we examine how we relate to reality. If we relate to technology as tools, devices, application, websites, and systems, we keep technology as something separate from us, something that could be a useful system or device one moment and a huge interruption the next. We must bring our attention away from one of devices and utility to an awareness of a different kind.

We need a technological space that supports, amplifies, and brings clarity to our relations with ourselves, others, and current issues. This environment would provide much more support in anticipation of what interests or concerns us. It will elevate our comfort and peacefulness.

Regarding episteme, I prefer a view of technological spaces. This view extends beyond more recent history. It also relates to our physical orientation. For instance, we rely on process coordination. I see us moving into a structural orientation. We are now entering this technological space, where centricity is no longer system or device oriented. Instead, systems, devices, and other technologies are tailored to fit each of us. What is important to us each moment is enhanced by the environmental shift or new technological space we live and work in.

Are you borrowing aspects from Foucault's modern episteme?

Thousands of years ago, systems were described as uniting and as bringing together. It wasn't until the 1800s when systems began to shape our understanding of the world we live in. This is a fairly young but dominant discourse, infiltrating almost every realm of our modern understanding.

Foucault speaks of this era as the modern episteme, and I also can anticipate that in the future this era might be referred to as the "system episteme," or a system technological space.

Yes, I am borrowing from earlier eras. As Foucault consistently illustrates, each era appears to have essential elements founding the next. This movement of thinking gives us a new realm or era we are now entering and exploring. This new era appears to have a relationship-based, structural, and centric orientation. It is being ushered in by the eras that preceded it.

Systems theory and its concepts are so engrained in our mind that defining the world as a complex system of interconnected elements simply isn't required. We view systems as closed entities that still coordinate with the environment. This rigidity suggests that the invention revealed in this book might be short lived (maybe a decade or two) but very necessary. With a systematic approach, it will create a bridge from simply using technological systems to dwelling in a system that relates with what is important in our world. Systemization must open up this new era before we can progress further with technology.

This invention, as a systematic approach, encourages the notion of systems as structures. The idea of interactive, interrelated, or interdependent elements forming a complex whole gives us an opportunity to eliminate process orientation. Interrelatedness provides structures that are highly functional.

New period of interest centricity

What about interest centricity?

How about if I cover centricity, interests, and concerns briefly, while leaving the design and methodology of centricity to chapter 6? Concerns come into play when interests are a source of trepidation or worry. In this new technological space, you gain a facility where your troubles and concerns can transform. Your actions align with your commitments, and—during turbulence or surprise—you gain access to greater privacy, magnificence, and trust through responsive networks of help and support.

Interests and concerns are a central focus of interactions. They shape the environment, creating a virtual room for all necessary supplies as you engage in interactions, adding value for others as well. The possibility of having new industries, departments, and other groups that provide value is a natural consequence of structuring our lives and workplace in this manner.

Recently, we have been engaging in social networking; a powerful social revolution seems to be in motion. It appears to be stabilizing, possibly signaling a conclusion to its rapid growth. Logically, the next will be a financial and economic revolution, due to the enormous economies, effectiveness, and profitable advantages available for individuals, families, and businesses. I foresee that organizational interests and concerns will fuel social involvement as our interests are being fulfilled.

The implications for businesses are for powerful, productive, and respectful relationships as we engage in rich, aware, and harmonious interactions. These conversations of splendid clarity give humanity and its institutions benefits that honor what is important in life.

Why is this design interest-centric and not another means for organizing ourselves?

You and I enter into and are motivated by specific interests. What interests us is important to us. We create specific projects and events and interact with what matters. Most times, we know that for each project or event that we take on, if we have certain people and resources in action, a successful completion is readily available.

A natural, powerful structure is one that keeps us in tune with our interests. Each interest will attract associated thinking, projects, and events. These occurrences might have subinterests and subprojects. This gives us a simple structure that is hierarchical in nature, completely displacing linear processes. These structures are produced and created in dialogues, in our daily conversations. These interactions and continuing dialogues evolve in time, from speculation to planning specific commitments in each specific area of importance.

Ultimately, if we reflect for just a moment, we can see that interests make up the core of our existence.

Why is that?

As children, our relationships with family and caregivers provide our earliest involvement with life. We listened to the patterns of words and contexts of interests and concerns, couched in the expressions of living life. These repetitive happenings solidified into understandings and meanings—coherent patterns that we recognize as thoughts. Thoughts are based in language, and language arises from the repetition of consensual interactions. Recurring conversations among families and friends reinforce familiarity and connectedness in these close relationships. This invention grants us a simple environment for what we were born into.

1. Martin Heidegger, *The Question Concerning Technology and Other Essays*, trans. William Lovitt (New York: Harper & Row, 1977), 3–35.

2. Michel Foucault, *The Order of Things: An Archaeology of the Human Sciences* (New York: Vintage Books, 1994).

Chapter 6
Technology, a
Methodological Look*

Structure displaces process orientation

Does this mean you aren't going to discuss functionality?

Yes, from a systems or a design parameter, it will be much more appropriate to reveal the structural aspects of the invention once it is revealed functionally in part 4, chapter 10. The notes that follow chapter 10 will provide you with functionality after you are aware of the alignment of the software, design, and structure. We can effectively go into structural and interest orientation, though.

What is a structural orientation?

A structural environment requires that communication no longer be a linear exchange of information. Information is no longer limited to passing from one person to another. Information and the activities that bolster it are not simply something to be stored in folders. The first crucial element of this environment, then, is the end of linear back-and-forth dealings. The second element to interrupt is an emphasis on storage and (isolated and secluded activities such as a) preparation for events. The third aspect to undermine deals with the management of elements like sales contracts, purchase orders, requests for proposals, and project management. One more aspect that detracts from simplicity is the elaborate array of checks and balances, with different processes for different people. Approval and procurement processes, where numerous people are involved in a single issue, are also part of the subject for reorientation.

It isn't that we don't know what a structural environment is; we do. We know they are effective, efficient, and economical. The concern is often whether they are consistent and coherent. The reconstruction of communication, accounting, and technology is required to eliminate this concern and provide a simple structural environment.

Communication occurs in a spatial environment; it evolves and changes. As things get planned and resolved, the environment gets simpler, not more complex. A structural orientation mirrors what occurs in the world and not what appears in the morass of "information storage." The only aspect that is linear or that exists in a linear way is how the structure of a calendar illustrates time. However, given our new particular access to the calendar, it isn't linear at all.

In part 3, chapter 7, we will observe how accounting also occurs in a spatial environment. As things are agreed and committed to, the future becomes more realistic. The only aspect of this spatial accounting that might be considered linear is that time is consistently associated with the interaction as it occurs. There are agreed upon times for commitments. All of these moments of time exist in the structure of a calendar in a structural way. Processes outside of production will literally disappear.

But what about approval processes, training, indoctrination, and so on?

I will go into this in more detail, but new employees will have more stringent, apprentice-style supervision. You have immediate awareness of plans and changes in plans, as well as the impact they will have on finances and resources. Transparency might be impossible in a process environment. It exists easily in a structured environment. The structure of the relationships in this environment becomes the structure of the organization. Nothing linear remains. I will go into this more later. Let's stay focused on this novel technological space, the advent of structures, and spatial relativity.

What is there to say about structures in this new technological space?

Structures of our own design are a significant part of the dwellings we reside in. With the invention, life becomes simple as it gently moves toward us, following our intentions, as we move from one interest to another during the day. This movement manifests in our interactions.

In this new world, fresh structures are constantly being built, and the structure in which communications happen is altered continuously. These alterations increase in detail and physical size as we coordinate with others and our environment. As we consider more opportunities and possible conditions are examined, the structure and permutations grow.

When we resolve and narrow our intentions, the structure contracts, and conditions of agreement narrow the design and content of the conversational structure. Certain parts of the discussion may disappear, erased (from the present, relegated to the past) because they are no longer relevant. Some parts might be sent into the future. At any time, interactions of the past can be reopened.

Our conversations facilitate the movement of these compositions over time, which brings us sense of change; as our interactive structure changes, communication, organization, and relationships progress. The structure also highlights what we are currently interested in and gives a vivid awareness of our current future.

Inside each interest, if we are concerned about the past or the future, that view comes into our awareness.

Design principles

Are there any basic principles this structure is based on?

There aren't really any rigid principles. However, there is a methodology (in which I took the liberty to reconstruct this new paradigm) for organizational design we can draw from. It is a natural thinking that was developed in creating a physical setting—the University of Oregon—that will work well in principle for a virtual setting also:

Specifically, we believe that the process of building and planning in a community will create an environment which meets human needs only if it follows six principles of implementation:

1. The principle of organic order.

2. The principle of participation.

3. The principle of [interest-centric] growth.

4. The principle of patterns.

5. The principle of [awareness].

6. The principle of coordination.

We recommend that [entities] which [have an organization], and [an autonomous financial structure], adopt these six principles to replace its conventional master planning and conventional budgetary procedures, to provide the administrative resources which will guarantee people the right to design their own places, and to set in motion the democratic processes which will ensure their flexible continuation.

For the sake of concreteness, and to give you an overview of the [methodology], we now outline these six principles.

1. The principle of organic order.
   Planning and construction will be guided by [structures and relationships,] which allows the whole to emerge gradually from [interactions].

2. The principle of participation.
   All decisions about what to build, and how to build it, will be in the hands of [each person, with appropriate relationships to maintain a coherent accountability].

3. The principle of [interest-centric] growth.
   The construction [of structure and design] undertaken in each [interaction] will be weighed overwhelmingly towards [interests, sub-interests, and] small projects.

4. The principle of patterns.
    All design and construction [of structure and design]
    will be guided by a collection of communally adopted
    planning principles called patterns.

5. The principle of [awareness].
    The well being of the whole will be protected by [a
    transparency afforded by structures and relationships
    – executive alerts – through interactivity accounting]
    which [reveals], in detail, which [interests and
    projects] are alive and which ones dead, at any
    given moment in the history [and future] of the
    community.

6. The principle of coordination.
    Finally, the slow emergence of organic order in the
    whole will be assured by a [financial and resource
    awareness] which regulates the stream of individual
    projects put forward by [those accountable].[1]

Interest-centric design

What is the significance of interest centricity?

The environment and every structure of support alter themselves based
on the current interest and circumstance. Outside this centricity, the world
seems complex. With the invention, what is important to you is made
obvious and placed inside a structure. Your interests and concerns are
continually apparent, and you live in an environment that provides a direct
access to simplicity, peacefulness, and purpose. The world we have been
living in and the way of thinking that gave us how we organize things is no
longer sufficient for living fully. Comfort and happiness dissipate quickly in
our current world, but a new world is possible. We have the technology.
We can build a different existence—one that is relevant to you always,
even as your attention shifts from one area to another.

Each interest or development has its own virtual room. The nearest virtual
companion opens as you approach to provide assistance. The centricity of
this technological space becomes available wherever you are; the nearest
device supports you. What that device provides depends on you. Each
space is your companion. You will no longer live in a device- or application-

centric world but rather an interest-centric world—a world of your current attention and intention.

The current interest pulls forth an environment from which to support present, past, and future spaces of interaction. The structure for the linear passage of time is in the form of a calendar. The interaction expands, contracts, and alters throughout its existence. Any point in time can be recovered and interacted with from the perspective of the present moment.

A hierarchical interest structure and its association with relationships and interactivity accounting create the environment, a strategic view of the future that has never existed before. Visibility and capacity, coupled with the efficiency of communication, open up the possibility of giving each interaction a weight of effectiveness only expected from the most experienced producers.

Looking at centricity in our inherited technological discourse, the necessity of moving from object to object has been causing difficulties for quite some time. This "spatial relativity technology" alters this discourse, having technology occur as a space, a place for you to dwell no matter where you are. We have a finite number of interests and a finite number of projects or developments, each with specific interactions, explicit resources, and precise elements of support.

How does this relate to people and resources?

We relate to people and resources in ways that are interest-centric. They are also project-centric and provide a structure that underlines the specificity of the interaction(s), as well as their historical involvement and evolution over time. People and resources are associated with a number of interests/projects.

The capacity to learn and anticipate during interactions occurs in a dynamic manner. When illustrated, the people and resources available vary, depending on the current project. Resources can be specific to a closed network, part of organizations, or part of the public network.

Resources provide a place for reoccurring items, solutions, places, public networks, and domains. Resources can also give entry for visitors, guests, advertising, and so forth. The upside for advertising is that it is context-centric, has viability of location, and is time appropriate. Advertising is a powerful and welcome benefit, rather than background noise.

To simplify our environment, ultimately people and resources will be addressed as resources. What gives immense influence for resources isn't simply interest-centricity; it is the added dimensionality relationships provide.

Say more about relationships and resources.

Relationships are the blocks used to build the structure and capabilities of this technological space. Relationships characterize what is important to us and reveal how helpful resources might be. Resources aren't just people, material, services, and advertising; they can be categories also. Since resources can be groups or collections, relationships can also be thought of in terms of memberships. These collections are made possible by entities invented within organizations. This will be revealed more succinctly as we continue.

Relationships, along with the design of the technological space, provide a vast awareness that is not ordinarily available to us. For instance, without this structure to discriminate, let's say we are thrown into a moment of disappointment. Without this capacity to discriminate, we might unilaterally dismiss resources across all domains. This might undermine our connection with others and our base of support. This environment, to a large extent, gives us a space from which to handle such moments. We react by removing only resources that are pertinent to the current failure in a specific area of interest and concern, whereas we can rely on that same resource in different areas where competency has been illustrated repetitively.

In order to gather a history of interactions and shifts in relationships before this invention, one dealt with a haphazard collection of documents, e-mails, and memories. With the invention, relationships and resources exist in a powerful structure that gives this historical account.

One important part of resources is the people you rely on—people that you know personally, professionally, or through public sites and that become a source of support for your current interests. There are also operational, institutional, and physical entities of all kinds that help us realize current and future plans.

Resources are usually sources that you can lean on when grappling with something important. They are introduced into your current interactions and removed when they no longer satisfy an interest. Some provide momentary support, while others provide research-related support.

Resources, both people and other sources, are made available explicitly in stuff. Practical resources include spreadsheets, word processors, purchasing templates, a GPS system, and a weather report. It is from our resources that benefits are derived.

In this invention, how do interests and stuff matter?

Our interests and concerns serve to change the space of awareness. Our interests might organize themselves into projects. While these take place, the space shifts, and the stuff correlates itself to the project, event, and area of interest.

Some of the stuff available to plan our activities and preferences moves us into different spaces, with the patterns of our activities providing for their design. For instance, between 6:00 and 6:15 p.m., we might view current events in our city as we drive home. We spend time with the kids between 6:30 and 7:00 p.m. and eat dinner at 7:00 p.m. A conference call occurs regularly at 8:00 p.m. on Thursdays, while Friday nights are date nights. Our lives are designed by interactions and reoccurring commitments determined by prior communications.

Interests and concerns are distinguished and identified from phrases, communication relativity, and preset concerns. They can also be distinguished within projects.

In this new world, honoring our interests and concerns, projects, events, and meetings allows for a powerful means to take life on and work in a precise way. Between moments of engagement, earlier interactions and conversations are immediately available. The time and effort lost in reconstructing where we left off in our thinking, research, conversations, or negotiations are regained. A linear continuance falls away. Stressful moments disappear when you are dwelling within your interests. What occurs as important to you, in life and at work, is now available each moment.

A simple coherent shift—process to structure

You mentioned eliminating processes. How is that done?

Business practices include a whole grouping of methods, processes, procedures, rules, and more to provide a coherent and pertinent effect for the commitments, goals, vision, objectives, and purpose that an organization holds.

All business practices can be reinvented from written rules, procedures, or processes and transformed into structural relationships. Rules, in a sense, are articulated and altered through interactions as trust and levels of competency are established. Business relationships are shaped for maximum efficiency, support (mentoring, coaching, and collaboration), and effectiveness as accountabilities are structurally reinforced.

Simply put, business practices are processes that originate within interactions. We have the opportunity to limit notions of processes to communication—specifically, to structurally redesign or reconstruct processes, procedures, practices, rules, measures, methods, and the like such that relationships for accountability are flexible and ever evolving. Practices take on creative design in speculative interactions. They occur in a very powerful and transparent environment with clarity and enhanced capacity for responsiveness.

In a process-oriented environment, there is a feature—"executive concern alerts"—that makes no sense at all. It creates group transparency, an amazingly powerful awareness for those accountable. I will cover this more fully in chapter 8.

1.   Christopher Alexander, Murray Silverstein, Shlomo Angel, Sara Ishikawa, and Denny Abrams, *The Oregon Experiment* (New York: Oxford University Press, 1975), 4–6.

Part 3
Interactivity Accounting: Powerful
Awareness, Scheme, and Structure

Chapter 7
What about the
Transformation of
Accounting?

A dimensional expansion

What is interactivity accounting, and how is it different?

Let's begin with current practices. What we have inherited, which must remain in the background for now (to maintain current continuity) is financial accounting. Financial accounting is the product of recent alterations of dual entry accounting, first captured by Luca Pacioli (as double-entry bookkeeping) over five hundred years ago. The notions of debit and credit prevail. Accountants gained their notoriety and certification just before 1900. New standards arose shortly after the Great Depression, and more recent codes have been standardized to reduce explanations.

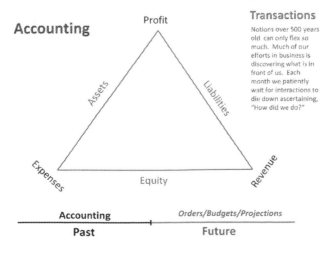

Accounting

Profit

Assets

Liabilities

Expenses

Equity

Revenue

Accounting   *Orders/Budgets/Projections*

Past   Future

Transactions

Notions over 500 years old can only flex so much. Much of our efforts in business is discovering what is in front of us. Each month we patiently wait for interactions to die down ascertaining, "How did we do?"

The simplest structure for accounting is represented by the following formulas:

Assets − Liabilities = Equity

and

Revenue − Expenses = Profit

Conventions like billing, invoicing, and resulting elements like accounts payable and receivable support accounting. Other conventions support these activities, including purchase and sales orders. Where these entry systems leave off, other systems have been created to fill in. I will mention these others with broad strokes:

- ✓ Management accounting

- ✓ Project management

- ✓ Customer relations management

These are a few of the many conventions. A newer means of compiling different sources is data mining all of these different entries. This data compilation is expressed in dashboards and similar displays of information that greatly assist management.

Interactivity accounting adds a dimension to accounting that can only occur at the moment the interactions take place. All transactions are initiated by interactions. There are natural mechanisms that have interactions appear as transactions so that current accounting practices aren't being violated (though the accountancy profession/society will likely drastically alter standards and practices in the next few years, given the power of interactivity accountability). What's unique about this shift is that the activities of entry processes that lie outside of interactions will no longer exist. Accounting entries and interactions occur concurrently in real time.

We will no longer need any processes like bills, invoices, orders, and contracts outside of the interactions themselves. This might seem a little abstract right now, but when we explore the third dimension of accounting, it will become clearer. The next level of clarity will come through specific examples of utilizing the invention in the next section. Ultimately, the highest clarity will arrive with the involvement of living life within the invention, as each of us design and are brought into our organizational environments. Nothing promotes awareness more than experience.

What is this third dimension?

Dual entry accounting can be viewed through the lens of time. Income—or revenue – expenses—occurs in periods of the past. By comparing the periods, we can view the movement and change in profitability of an organization. Though the balance sheet is also reported each period, it includes a current- or comprehensive-looking aspect of the condition of the organization as of that period in time.

Comparing earlier balance sheet periods provides a movement of organizational conditions historically. A balance sheet is the momentarily then-current status of equity—or assets – liabilities. We can view the change of property, wealth, or value. When looking at organizations from this perspective, we are less aware of the profits or losses in any given period and are more in tune with changes in the fundamental value of an organization.

Another basic view of an organization is in terms of cash or cash flow. We can anticipate trends of the sources to cash through operations and/ or finances. There are numerous other views that measure projects and manufacturing. These views are resource commitment driven. The number of different, useful measures varies greatly based on unique contexts.

For those who don't have a background in accounting, these conventions— income, balance, and cash flow statements—might sound convoluted. These practices have been needed, given the current paradigm of accounting. This contextual shift of accounting might empower the world of accountancy to create something very creative and very simple in the next few years.

Initially, a transition will be required. The third dimension is what occurs during most interactions. Much of the time, it relates to the present and future. One way to keep this current aspect of accounting coherent with the present is to maintain ideas of current and future equity. What opens up in an obvious way is a benefit of a third dimension of accounting in terms of value—or (equal to) agreements – commitments. We are taking the entire notion of accounting into the future as only accounting for moment-to-moment interactions can allow.

71

What do the terms agreement, commitment, and value specifically mean?

Agreements, from an accounting and legal perspective, are binding mutual contracts. The definition itself relates more to an act or an arrangement for a future course of action. The use of the term "agreement" is more closely related to the latter concept.

Agreement pertains to offers and sales agreements. More specifically, it is the portion of a contract or mutual promise which, when the interaction transacts, occurs in a future time period as revenue. From a dual entry accounting perspective, a term like future receipt or an implied term like receivable for a future time period would also fit a notion of future assets.

A commitment can be thought of in many ways. It might be a promise or pledge to do or not do something. It also refers to a devotion or social devotion to an ideal.

From a financial perspective, commitment means assuming financial obligation. Expanding this further, a commitment is a future plan involving resources and money. It might equate to a future expense.

Buying, employing someone, renting, and scheduling utilities or services all require a commitment and exist in a committed structure.

What I conceive of to temporarily maintain a coherency with current practices is to permit a relationship in time between an explicit agreement of sales and receipts and commitments of purchases and payables to be implicitly regarded as dual entries. In the same vein, the time periods between sale, purchase, delivery, payment, and receipt give interactivity accounting a structured form, where transitions are demonstrated in time, empowering the abandonment of process thinking. The future difference of agreements and commitment imply value. As the future arrives, these interactions contain resolutions in which what was previously related to as something transacted, now simply implies profit as the conditions of value are fulfilled (complete).

The formula that represents a future span of time is written as Agreements − Commitments − Value = Assets − Liabilities − Equity, while the past period continues to be implied by Revenue − Expenses = Profit.

Now you can begin to see what had me recently say, "We are taking the entire notion of accounting into the future as only accounting for moment-to-moment interactions can allow." We have an immediate balance sheet in each moment (now and into the future) based on the current state of all interactions, and we are aware of how every interaction alters the current future.

We can view the current future status of now, ten minutes from now, the end of the day, one week from now, two weeks, a month, two months, and so on. The current rate of growth and its changes are constantly viewable as well. The benefit of the third dimension for accountants, in terms of value or agreements – commitments—is obvious. For management, a whole new world opens up. Awareness of value, the management and fulfillment of conditions or agreements, the contracts and changes and commitments, and the effective use of resources and funds committed all provide a vivid future every moment.

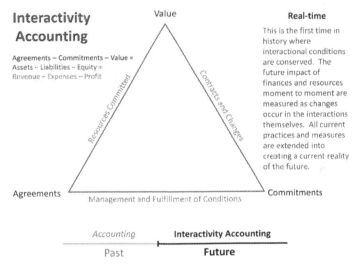

What about when there are changes or delays?

That is a perfect question because, until now, projections and budgets have frequently been thought of as suspect. It's often reasoned that the thinking behind these plans could be too optimistic, too pessimistic, or biased in another way. The best CFOs and CEOs reach success because they have a way of anticipating or sensing the direction and quantity of disruptions that could take place. Organizations spend a tremendous amount of money and resources to gain a sense of the future.

73

The money and resources spent capturing the present and anticipating the future will no longer be necessary. In fact, if you can imagine this, the current future is always visible. When changes or delays take place, their influence on the finances and resources of the then-current future are immediately visible, in the moment of that interaction. The change of the future becomes apparent. I think this will excite organizations greatly. This transparency can be general or very specific as a design function for organizational accountabilities.

Accounting—a design function

What do you mean by a "design function" for organizations?

Remember when we spoke about how this new environment might limit personal autonomy at junctures where work and the interests of the organization coincide? This now becomes a design function for the organization.

Business rules and the management's awareness of their area of accountability become transparent. Depending on the desires of the manager(s), the interactions that might influence future resources, deliveries, and finances become visible. They are not visible as information but as the way in which the new current future influences finances, commitments, and resources. This is important because this environment doesn't deal with information. As a manager, you are now part of the interaction. You view the interaction and the current or recent changes.

> We are searching for some kind of harmony between two intangibles: a form which we have not yet designed and a context which we cannot properly describe.
> —Christopher Alexander[1]

Another way to comprehend this shift is to recognize that information requires processes. Structural design (management, employee, and third-party relationships with resources—the relational design for awareness of plans, changes, and interactions) has specific advantages when developments are conversational, interactive, and relevant.

The opportunity for managers to become mentors expands tremendously. The employee is no longer dealing with issues using only his or her own capabilities and awareness. The weight of experience of the manager is now available as well, alleviating many delays and improving on original negotiated conditions. Resource allocation and profitability improve. The "problem" can, at times, transform into a new advantage or benefit. This becomes available through a natural and powerful means of problem solving, impacted by expanded and more effective means for speculation and adaptation.

How will this expand our effectiveness in speculation?

You have complete control of the now-current future. As you speculate, the speculative future immediately comes into view. The way that resources, cash, and profits could shift is discovered when looking from a speculative view. Can you imagine how this capability will alter an interaction in strategy and planning? Remember, nothing takes place outside of this interaction. There is no waiting for reports or analysis. The current future is immediately available, each moment.

All of the creativity and focus needed for a specific and beneficial view are inherently available. Job costing is no longer a process; it is real-time, altering as interactions occur. Projects are planned in an interactive space. Project management is simply viewing projects as aspects and expressions (conversations) of specific interests.

How will projects be organized?

In order to get a sense of this, it is important to dismiss any notion of linear communications. A project is simply an interaction (virtual work area) that continually changes: beginning, increasing, and then decreasing in complexity as plans are initiated, changed, and completed. Interactions as they occur are relative to the current moment viewed. I will have to illuminate the calendar function, or interactivity scheduling, for you to get a sense to the immense clarity and power of this interactional work area (space).

What is interactivity scheduling?

Many portions of planning and interactivities are date sensitive. Unlike calendars and current scheduling applications, this form of scheduling relies on expressions in the interactions. Once available dates and times are found, a meeting or reconvening (choices that fit each participant's current future) is suggested and chosen (warnings and alerts are put into place also). Unlike typical calendars, there isn't an entry in a calendar; the interaction occurs there. There aren't resources or finances scheduled; they are allocated—as discussed and agreed—in time, in the current future.

Resources, finances, agreements, commitments, terms, and such are displayed in a calendar-type structure along with interactions, events, projects, and meetings. Access to calendars and schedules provides an illustration of changes in the current future, which reveals the effects of current or recent interactions. A historical rewinding is available within the calendar as well. Selecting a specific time provides an awareness of that current period and the changes that occur each moment.

Each interaction is expressed over time in the calendar. When expressions are no longer valid or applicable, they are removed from the current interaction. Future events can be planned and removed (and slipped into a future period in the schedule, literally dismissed) from current relevancy. Regardless of when they occur, interactions (or a specific state of an interaction) are designed to show up only when they become relevant.

Readily apparent is the fluid nature of interactions. Within interactions, a discussion might preempt future accounting issues and changes to the schedule. These issues might occur earlier than planned. The current future remains coherent because the calendar serves as a contrivance for scheduling financial occurrences, as well as resource allocation.

Both scheduled plans involving agreements, commitments, and resources and the moments when interactions are taking place are relevant to the now-current future. Potential views of future periods help measure changes in value. Differences in the now-current future are significant because they illuminate how agreements and commitments transform during an interaction. Over time, the changes/transformations occur in the same interactive space. By visiting an earlier moment of time, the interactive space would reveal the condition of the past interactive space.

Let's say that a widget is purchased on January 23, scheduled for delivery on January 27, and payment is scheduled for February 10. The interaction and accounting occur and are available on January 23, January 27, and February 10. The interaction itself is available on the calendar, and if the delivery is delayed on the twenty-sixth the interaction is updated so that a January

28 delivery will occur instead, the interaction and delivery shift from the twenty-seventh to the twenty-eighth. By revisiting the interaction on the twenty-third, we see the conditions prior to the delivery change. From the twenty-sixth on, new conditions exist. This is interactivity scheduling. It is the very structure that interactivity accounting exists within.

The current future

What are the current futures and the now-current futures?

New current futures arise amid the interactions or collections of interactions that take place between individuals, in families, or in organizations. The interactions continuously address the future (or the past) in such a way that the future gains clarity. From a financial or management perspective, this notion is incredibly powerful. This term centers on the idea that the future moves and changes over time. Each interaction could introduce new speculations, agreements, or commitments, and it might alter the consequences of previous conversations, agreements, commitments, and terms in an interaction (or interactional space—virtual whiteboard).

If we are ever curious about a previous current future, the calendar permits us to go back in time. Prior agreements, commitments, and speculations are retained chronologically. In the same vein, future aspects of the current interaction might not appear until a later, appropriate time. They will appear when they become relevant to the current interest in an interaction. If, say, you've scheduled a delivery, indications regarding this delivery will appear closer to the day of.

The historical and future relevance of an interaction is accounted for by interactivity accounting. Changes in agreements and commitments that alter our speculations are accounted for within the structure of interactions.

We are ready to begin the next chapter.

Can you quickly recap where we are?

This book introduces patents for an invention. The invention brings us into a new world, a world where our immediate interests and concerns are supported and accounted for. The support lends itself to an environment that is completely in tune with the current conversation. It is like a virtual reality mapping directly onto our physical reality and offering support like people and resources to help us. Tools and applications are designed to support our actions. When we think, talk, or listen, we are aware of our emotional state. We are aware of whether we are engaged, effective, connected, and aware—or not.

We no longer have to learn how to use applications. We no longer need to take additional actions outside of our conversations and interactions. We no longer need to capture data, schedule events, cover expenses, or any other processes that come to mind, outside of the moment of an interaction. Furthermore, we no longer have to reacclimate, or find where we left off, in regard to initiatives. All we have is our environment, and we get to design it as we want, based on what is important to us. Wherever we go, our environment is with us. We get to control how our environment works when we are in the car, walking, in our homes, at work, or in a public place. Our environment, our virtual dwelling, shows up wherever we are.

Last question. How does it show up where we are?

Initially, our virtual reality might be available for us via a mobile device or signal that lets the devices closest to us know that we are present and our environment is required. The screens that are closest to us provide us with a consistent environment, coherent and in line with our current interests and interactions. A thumbprint or other means might be required for authentication if there is any question of identity. One day (maybe very soon), authorization could take the shape of an optical scan or facial recognition—use your imagination. There are many ways to accomplish the reality that when we show up, so does our environment. The devices can be coordinated by the combined recognition of us and our location, a function of entity support that we will discuss in part 4.

1. Christopher Alexander, *Notes on the Synthesis of Form* (Cambridge, MA: Harvard University Press, 1964), 26.

CHAPTER 8
ACCOUNTING: AN ONTOLOGICAL AND STRUCTURAL VIEW*

A break from antiquity

What about accounting?

Accounting is thousands of years old. The earliest accounting records, dating back more than seven thousand years, were found in Mesopotamia.[1] The peoples of that time relied on primitive accounting methods to record the growth of crops and herds. Accounting has evolved, improving over the years and advancing as business advances.

Today's accounting is very complex. The amount of resources that organizations consume to achieve a degree of accuracy and certainty is daunting. In accounting, results are assigned to meaningful categories and distinctions for further analysis and evaluation. Temporality is an obvious component, as dates play an inescapable role in accounting. Recording changes, analysis, and reporting what we expect all reflect its temporal quality.

Interactivity accounting introduces a paradigm shift. Before considering this shift, it might be instructive to consider accounting as a structure for interests and concerns. Without shifting anything but our view, we expand the possible scope and reach of accounting as a structure that supports everything important. The introduction of this new paradigm is based on the notion that what ultimately gets captured must originally occur as an interaction, even if that interaction is simply between ourselves and our thoughts. An interaction can be as simple as a speculative curiosity.

I am interested in expanding our understanding into a language of organizations, where finances, resources, and time commitments are accounted for. The desire is to expand beyond traditional thinking, such as bookkeeping, auditing, management accounting, analysis, inventory, and costing. The opportunity lies in expanding the scope for management of schedules, agreements, commitments, and even speculative ventures, estimations, formulations, and strategies.

What makes interactivity accounting unique?

As a prerequisite, the earliest evidence of double-entry bookkeeping was the Farolfi ledger of 1299–1300. Luca Pacioli, though not the inventor of double-entry bookkeeping, wrote a paper on bookkeeping that laid the foundation for double-entry bookkeeping as it is practiced today.

This current methodology captures or declares transactions for completion. All the practices used to manage projects, sales, and purchases for the future are complex attempts to manage activities and commitments. There is a fundamental blindness in our current tracking of these activities.

Our view of the future is increasingly important, and businesses expend a lot of time, money, and energy to better anticipate the future. We are introducing a new, reliable model that opens up this world. Each of us, as part of families and organizations, has issues that require our attention and require measure and accountability.

Professionals who are steeped in the world of accountancy view accounting as an entry process of measuring something that already exists. This pull for existence that is material might immediately negate a trust for accounting for the future. It appears that often people are too optimistic, but perhaps the breakdown isn't in optimism. The degree of promises or agreements (and/or commitments) made or not made, kept or not kept, might also lead to this distrust.

A novel notion expands accounting fully into the future, adding a dimension to dual entry accounting (and double-entry bookkeeping) while at the same time dissolving the notion of posterior entries. It also adds a simple means of expanding analysis, vision, acuity, and scrutiny in the current future. This relates to communication and accounting about speculation, agreements, and commitments as they occur in interactions. The pull for materiality is satisfied in this coherent structured environment.

Interactivity accounting correlates directly with time commitments for finances and resources. The balance sheet (i.e., Assets – Liabilities = Equity) is related to in the present, and future may be divided into various periods or sections of time as related to a future goal. This goal (or area of accountability) might occur as a value statement, or quasi ledger, in which economic interests are represented as Agreements – Commitments = Value.

These representations of measurement can be manipulated quite easily for management of projects, departments, and other divisions. The changes in state between two moments, on the other hand, create immediate awareness from a feature called "executive concern alert." From an organizational perspective, changes in value might trigger openings into management conversations for a business or alert a parent of difficulties or change of plans within a family.

The elements of interactivity accounting take a makeup of identified notions like services and products into consideration. These notions are part of interactions and templates throughout time.

"Value" and "future equity" are components having two distinct temporal aspects. These constituents distinguish current future relativities, akin to the starting and ending differentiations in any measurement of time. Some existing conventions illustrate this basic concept. An order, for example, might have an order date and a delivery date and alert one to changes in these dates. This concept of having a time and date, as well as terms of agreement, for an interaction is certainly not foreign. We recognize that our interactivities create both potential and actual agreements and commitments.

What gives us the present current future from a perspective of value might vary from previous current futures because something has changed or been committed (and/or agreed) to between the past and now. Changes arise inside the interaction with circumstances, opportunities, and disappointments. Major changes could influence the value and/or the agreement/commitment date.

Accountability clarity and transparency

What is executive concern alert?

In a process-oriented world, buying something, planning something, introducing or changing something, and dealing with something outside your authority often require approval or notification. Several layers are generally involved in this process, and the possibility of initiating such a process requires a certain length of employment and experience.

In a relative spatial environment, executive concern alerts provide a body of relationships to finances, resources, and people, woven in the fabric of interactions and interactivity. A new employee gains different capabilities as his or her relationships with professionals, supervisors, mentors, or coaches shift. A shift toward greater autonomy comes after competency, gained from daily experience, is assessed.

There are several arenas in which executives gain access to interactions and conversations that are directly or indirectly connected to their accountability and scope of responsibility. This access is structure oriented and not process oriented. The relationship with people, resources, and finances provides a platform and defines this phenomenal transparency and awareness. Let us consider the following examples:

Finances—There are many areas where changes in current future or speculative changes in finances open up interactions for an executive. A few of those areas are dependent on department, region, product line, and project and event planning. One example for a project manager is if the current future value decreases (or increases) at any specific date by 0.5 percent (or any designated level of deviation). When that occurs, the executive enters the related interaction with a level of awareness about the parties involved that is defined by the relationship between the manager and employee.

Resources—There are areas where changes in the current future or speculative changes in resources or resource allocation also open up interactions for an executive. Those areas might also be dependent on department, region, product line, and project/event planning.

People—There are many areas where certain ideas, expressions, and phrases expressed within an interaction open these interactions for executives, accountants, lawyers, technicians, and others. An example of this is illustrated by the word "delay," "legal," or "postpone"—key phrases that alert professionals or other support of something that might be a disappointment or breakdown are part of this area of benefit.

Professional—From a human rights perspective, issues of privacy might be required to be distinguished. Privacy is anticipated in activities of the legal profession, and organizations must be responsible for this transparency. Security and intrusion are vital concerns in this area. These matters, for countries that have a democratic or libertarian foundation, are dealt with through legislation and by judicial means. For more authoritative societies, there will be a similar balance when organizations and alliances are nurtured in this new world.

I know you covered this, but where do processes go?

Processes disappear because, in time, the movement and flow of all interactions is time oriented, not information oriented, as inherited. This structure (interactivity scheduling) gives us a world where we can move backward or forward in time quite simply and where planning is an interaction placed in time as opposed to a series of activities. The relationship between persons and resources keeps plans coherent and consistent. Like everything else in this new world, these relationships are derived from interactions.

1. "Ancient Mesopotamian Accounting and Human Cognitive Evolution," last modified January 6, 2005, accessed May 26, 2012, http://www.redorbit.com/news/ science/116691/ancient_ mesopotamian_accounting_and_ human_cognitive_evolution/.

CHAPTER 9
ACCOUNTING: A TECHNICAL VIEW*

Accounting—structural orientation

In chapter 7, you mentioned patterns of accountability when moving into structural awareness. Can you elaborate?

To illustrate these advantages empowering real-time transparency, I've created an outline which allows you to view this new technological space in terms of features and distinctions.

- ✓ Patterns—Two examples:

    - o   It is relationships that determine a system's essential organizational characteristics and design,[1] and it is patterns that give rise to what we can identify (biologically).

    - o   Alexander provides another perspective (architecture): "We may define a pattern as any general planning principle, which states a clear problem that may occur repeatedly in the environment, states the range of contexts in which this problem will occur, and gives the general features required by all [structures] or plans which will solve this problem."[2]

- ✓ Pattern Language—Structured method of specific design patterns for accountability (resources/finances) and interests (projects/departmental responsibilities). They provide a nontechnical vocabulary for design. Each person has a pattern language that is "slightly different from the language in the next person's mind; no two are exactly alike; yet many patterns, and fragments of pattern languages are also shared."[3]

✓ Structural Cybernetics ("Cybernetics"—Greek, meaning "the art of steering")—this invention:

o Spatial Relativity Technology—Each entity (person or organization) has at its disposal each interest and concern as a relative environment

- Interactivity Interests and Concerns—Structural means to learn and capture interests and concerns expressed within interactions

- Executive Concern Alerts—Structural means to design transparency for those accountable for specific interests and concerns

o Communication Relativity—Feedback of current emotional/communication state based on a new communication model

- Interactivity Awareness—A means to differentiate bodily changes

- Interactivity Awareness Support—A means to learn language, inflections, and tonality

o Interactivity Accounting—A new structural accounting for resources, finances, commitments, and agreements real-time and effectively into the future

o Interactivity Scheduling—An integrated reflection of Interactivity Accounting and a temporal framework for interactions

o Interactivity Processes—Where interactivities are supported structurally

o Interactivity Interconnection—Structural translation for system support for nonspatial occurrences

What is structural cybernetics?

One view of cybernetics lies in the thinking of anthropologists like Gregory Bateson and Margaret Mead and neurobiologists like Humberto Maturana and Francisco Varela. Their theories share threads with the communication model, especially with studies of the feedback of emotions. The distinction "structural cybernetics" wasn't viewed through this particular lens, however.

My interest is in building on the technology of artificial intelligence. While some sense of a structure might exist for intelligence, with the capability to initiate environments that allow for intellectual coordination, cybernetics is not at all artificial. The structure learns and gains an ever-growing capacity to anticipate patterns for actions, interactions, and propensities of the entity that it structurally coordinates with.

What are interactivity processes?

A simple structure allows for interconnection. This structure includes participants, timelines, templates, borrowed elements, and other elements. The only process that takes place is the interaction itself. Structures based on relationships eliminate the posterior processes that plague both organizations and people.

This name of this feature is somewhat misleading because processes literally disappear. Structures displace processes. Relationships give structures for design.

A simple way to identify structures and templates is by their opening of portals to one or more systems. The schema, or translation, of these interconnections between interactions creates databases supporting interactivity accounting and awareness of resources as the content of interactions change.

This environment does not discourage systems or the logical progression of extremely viable enhancements for organizations. It simply discourages inconsistent coordination in life and allows for visual awareness of system enhancements within the structure of this virtual environment (as opposed to in some separate screen or environment). The faster organizations develop this transparency and interconnection with the environment, the more viable their systems will be in the marketplace.

Functional design of interactions

You have been talking about this new accounting methodology. If there are rules associated with this accounting, would you share a few?

Each interaction is related to an interest and potentially a project. Each moment that the interaction is altering, it is captured in time. On the screen, the calendar displays the then-current state of the interaction. Visiting or touching the calendar brings that current time frame of the interaction in the communication window.

All speculations, commitments, agreements, and other conditions are captured in their appropriate moments of time of the then-current interaction. As speculations, commitments, agreements, and conditions change, so do the appropriate captured moments in the future.

These changes in time, finances, and resources between the two periods of time of the interaction are apparent, and the changes in the financial and resource positions are accounted for. In other words, changes as they influence the future are readily apparent and visible to you and those accountable via the relationship with you, this interest, and project or event.

There are explicit and implicit influences for any finance or resource agreement or commitment. These resources might be groups, people, organization, or physical resources. The relationships for these resources and financial instances are defined (within interactions consisting of those persons responsible for the design).

The future aspect of accounting occurs in a structure defined as Agreements − Commitments − Value = (Future) Assets − Liabilities − Equity. Once completed, or when the interaction is transacted, agreements occur as revenues, commitments occur as expenses, and value occurs as profit.

Agreements and commitments occur and alter within interactions. Value, assets, liabilities, and equity are many times implicit, based on the relationships of the resources (and the subsequent agreements and commitments) within the interactions. The increment of time between periods is flexible, providing an immediate awareness of a change in value.

Depending on the relationship (accountability) with individuals or groups (of individuals), an individual's (manager's, mentor's, executive's, and so on) direct awareness and (transparent) entry into an interaction is immediate and readily available:

- Whether the participants of the current interaction are aware of the entry of an executive into an interaction is based on the relationship;

- The relationships are negotiated. The supervision for new employees might be extensive, allowing hands-on mentoring and training as opposed to a process-oriented training (and/or education);

- Accountabilities and organization structures are dependent on the relationships. As competency develops, the relationships change.

The portions of interactions that are future based or relative in the future become present when relevant. Only aspects of complex interactions are visible when they are relevant to the current interaction. What plans, speculation, agreements, and commitments that have passed or will be relevant in the future reside in the calendar/schedule with the capacity to move forward or backward in the calendar for any interaction gives what was or will be relevant

- in the interactive space and

- for the resources and finances, speculated, agreed, and committed to.

The immediate is all that is present. Accounting, time, and resources are simply what occur in interactions. The value of this difference between current agreements and commitments and the third dimension of interactivity accounting will result in a historical shift and an altered future. Whereas conventional accounting uses comparisons between periods, interactivity accounting provides awareness of incremental changes. This allows for a creative array of designed business and management practices.

1. Humberto R. Maturana and Francisco J. Varela, *The Tree of Knowledge: The Biological Roots of Understanding* (Boston: Shambhala Publications, 1987).

2. Christopher Alexander, Murray Silverstein, Shlomo Angel, Sara Ishikawa, and Denny Abrams, *The Oregon Experiment* (New York: Oxford University Press, 1975), 101.

3. Christopher Alexander, *The Timeless Way of Building* (New York: Oxford University Press, 1979), 203.

Part 4
The Invention

CHAPTER 10
A NEW WORLD BROUGHT ALIVE

A look at the design

A picture is worth a thousand words. Can you start off by providing a picture?

Of course I can. But I want to emphasize that the way this environment looks is personal—entirely up to each person. For instance, even though organizations influence the context of areas like accountability and management that you might interact with or intervene to a small extent in how specific areas of your environment look and feel, the terms utilized are yours to choose—no one else's.

> The structure of life I have described in buildings—the structure which I believe to be objective—is deeply and inextricably connected with the human person, and with the innermost nature of human feeling.
> —Christopher Alexander[1]

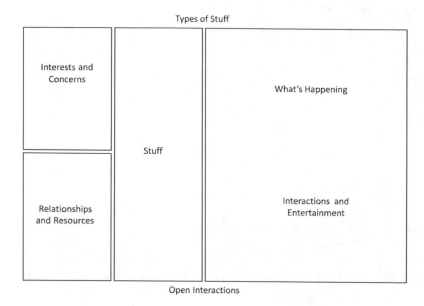

Now, first of all, it looks like the screen is divided into four sections. That is not at all required. If you want four sections, this is one possible way to organize it. Another option is to have the current "Interests and Concerns" in the top left corner where "Types of Stuff" is, possibly as a dropdown so that it won't take up space. Other interests might be on the right edge. You might want "Relationships and Resources" on the left edge, and pictures or videos could be the depiction of people and resources. I would keep the two areas for "Stuff" and "Interactions and Entertainment," minimizing "Stuff" if I don't need it like when I'm watching a movie or reading a book.

I will relate to this diagram for now in discussing what you want to know. Firstly, the "Interests and Concerns" might list all of your interests, but it also indicates what the main interest occurring in the moment is. Although you can manually select an interest, the context of the interaction, including where you are, allows the invention to decide on the current interest, which, in turn, manages the environment, resources, and stuff.

As the interest changes, the "Relationships and Resources" and "Stuff" changes too. There are several "Types of Stuff" for each interest, and the types change in correlation with your interests. I won't go into controls and settings except to say that you get to create different settings and controls, modifying how things look and work, based on personal preferences. The type of device used, your location, who can intrude when you're busy, how you are viewed, how you communicate, how your emotional state is indicated, and other such settings are all under your control.

The first thing that sounds obscure is "Types of Stuff." What is that?

You can think of stuff as the furnishings and supporting elements of the virtual room. Stuff includes tools, applications, lists, definitions, translations, templates, and calendars—all of which change based on the current interest and support specific purposes based on your location and activities. In the kitchen, it might provide cookbooks, stovetop and oven controls, and timers. If a recipe is written using metrics but your utensils are of the English system, it converts everything for you. I think you are getting a good idea of its uses from this analogy.

There are many things encompassed by "Stuff" and the different types of stuff employed are really up to you. There are some common pieces of "Stuff" that I can reveal and discuss, though.

- ✓ Calendars (Scheduling time, finances, resources, controlling interactions and interactivity accounting, and anything else in time)

- ✓ Templates (Recurring formations for tools, applications, conditions, finances, agreements, and so forth)

- ✓ Systems (Translation for applications and systems, residing on the web or locally)

- ✓ Interests or concerns (Domains, projects or activities, records—history)

- ✓ Relationships (Words, situations, and more are associated so that specific personal or professional relationships, resources, and support appear)

- ✓ Communication utilities (Reoccurring capturing of phrases/ emotional distinctions)

- ✓ Lists (of interactions, projects, entertainment, and the like)

Templates might be the simplest to think about, as they are instrumental— frequently appearing in or dragged into conversations. Templates can be thought of as built-in support. If you are in the car, GPS and traffic alert bots are examples of two templates or applications that might support you in where you are going. These applications would be tied into your interaction insomuch as there is nothing to enter; the GPS and other supporters assist you as you drive. This is an example in which you aren't accessing an application; you simply have plans and are in the car while the GPS assists you.

The "Stuff" adjusts itself and moves into your interactions to support you and your activities. As your interactions travel from one interest to another, stuff that is relevant to the new interest or activity becomes available and moves with you. As repetitive patterns are noticed, given recurring interests, activities, and situations, relationships are learned and retained.

Another way of envisioning this capacity lies within the interest of entertainment. Stuff is seen in a television list of programming or genres of programming and in the controls for the screen or other screens nearby. All notions of controlling your environment, especially when a screen is involved, are an immediate part of your virtual environment (stuff). All separate controls like television remotes and services like cable and satellite control are duplicated nuisances and lose all value in your life.

In time, all your lighting, appliances, and other systems will not require any controls outside of your virtual environment. This isn't the main purpose of this invention, but it is a pleasant side effect. Energy savings abound when smart applications are made available. These are just a few of the thousands of changes that will likely occur in our lives.

New industries spawning from this new world are beyond comprehension. "Stuff" expanding this virtual world will be an opportunity for third-party developers. As interests and concerns become prominent areas of life and work, rich entrepreneurial opportunities will arise.

This seems very complex. How could this possibly be designed effectively?

I really appreciate you asking this question. My response is that the design that makes this virtual environment simple to orchestrate was thought up years ago. Now that cloud and seamless networking are readily available and prevalent, the design simply involves coordination between entity domains and interactive (occurrence) domains—a concept incorporated into most e-mail, server, and online environments.

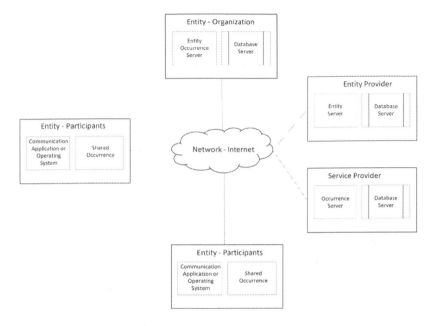

The figure above is a replication from the one of the patents, and it illustrates this concept, as there are two domains being provided:

- ✓ One that learns about the entity: the wants, preferences, expressions, networks of support—the people, resources, interests, concerns, affiliations, structures, and the like

- ✓ One that provides a shared environment for interactions, media, collaboration, agreements, and so forth

This simple design empowers the dance of entities/interests/support within a world of relating, interacting, and communicating.

Seeing it in action

This is too explanatory. Can you liven this up a bit?

Let's continue with the story of Bill and Alice's life that we visited in the introduction, only it is more than six months later. Bill's company, a prominent advertising enterprise, initiated a change from e-mail, applications, and a whole sundry of systems to this new technology that has become the latest rage.

Before we begin, I ask you to allow yourself to be swept into this story. You might have the experience you are being transported into the future. You will be! Not a future a hundred years from now, rather within a year from *now*. I ask you to consider that science fiction is fiction only when we don't have the technology. There is no reach; we have the capacity to do business and live in the new world you are about to see. Before we get started, keep in mind the explanation of the technology, communication, and accounting paradigm shifts you have been engaging in.

> Bill is viewing and interacting with his screen, an iPad loaded with a software system that has hijacked every screen he comes in contact with these days. Unlike what you or I might expect, the screen doesn't have a lot of icons. Instead, on the edge of the left side is "People and Resources." On the right edge is "Interests and Concerns." Currently hidden by the keyboard are different templates—drawing, writing, and other elements that, when the keyboard is retracted, get dragged into the largest area of the screen.

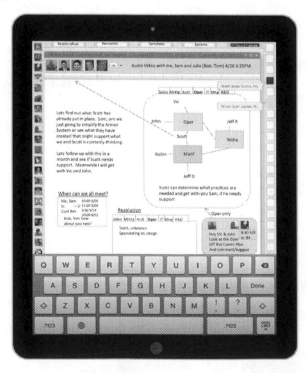

Here is Bill on his tablet, completing a marketing strategy meeting with Sam and Julia. Sam is Bill's boss, senior vice president of the eastern advertising division. He is very pleasant and easy to talk to. Julia and Bill have a history together. This will be the fourth joint venture they have done together in the last year. Sam and Julia are live, speaking and reacting as their video and audio are taking place right now. Bill is viewing himself, Sam, and Julia engaging in an audio/video conference where he and the others have full access to an electronic whiteboard.

Bill says, "Julia ... Sam and I need to get with some people here at Stanton Advertising to make sure we can manufacture what you need in June."

Sam says confidently, "I am pretty sure that we can take care of everything. I think I saw the stock in the warehouse last week, so I doubt we will need to order it. When should we get back together and start finalizing everything?"

Bill and Sam are producing a diagram on the right side of the virtual space during the conference. Together, they are organizing coordination with Vic and John in operations, Robin and Jeff Davis in manufacturing, and Jeff Ritter in the warehouse. Bill has worked closely with both John and Robin before, and things turned out well in the past. Scott is new to the company, hired by Sam to coordinate operations and manufacturing more effectively. Bill has never had dealings with Scott and is a little unsure of how they will work together. Since Sam hired him, he doesn't want to express his reluctance. Since Scott isn't available right now and he is needed to help plan the design, there are a few private areas and separate interactions set up, as Bill and Sam currently have full access to the electronic whiteboard. Having completed the diagram, Sam must have brought in a meeting template.

Bill smiles when he sees that Sam has included Bob and Tom. Both Bob and Tom intrigue Bill. They are both very young. Bob is a Harvard graduate and is as sharp as they come. Tom went to Bill's alma mater and has a lot of gumption. He can tell that both Bob and Tom will be heading up something important in the company very soon. Every dealing Bill has had with either of them has been very much a learning experience. The calendar gives four different dates and times when Bill, Sam, Julia, Bob, and Tom are available: May 5 at 10:00 a.m., May 8 at 11:00 a.m., May 14 at 3:30 p.m., and June 1 at 10:00 a.m.

Julia suggests, "How about the eighth?"

Bill and Sam agree. Sam says, "Thanks, Julia, we'll visit in a couple of weeks." As Julia drops off, Sam says to Bill, "That went well. It looks like Scott is tied up for a few days. We will need him to make this happen."

Bill suggests, "I can handle it!" Bill spoke impulsively, realizing that he was uncomfortable that he really doesn't have a track record with Scott.

Bill notices that his screen has taken on an orange hue. He realizes that he is a little arrogant and adds, "Maybe I could check with Mark and get his support." The screen brightens and shifts from orange to a slight violet-blue hue, one that he prefers. Bill thinks, *Mark gets along with everyone.*

Bill sees that he can get with Vic and John an hour and a half before they visit with Julia again, and he captures that time with them. He has already decided that he will direct everything through John. He is comfortable with John and knows that he and Robin work well together, just in case Scott doesn't work out so well.

Sam is satisfied that everything is in order and says, "Bill, everything looks good. If I can't get with Scott in the next few hours, I will have you notified. Get with Mark and see if he can support you. Add him to the 'Resolution' discussion, all right?"

Bill responds, "Thanks for your support on this, Sam." They disconnect.

What is going on behind the scenes is the two people not engaged with Bill, Sam, and Julia will view this meeting in the near future; this strategy session will continue over time and possibly develop into a project and a new division. Bob and Tom, who were absent from this interaction, will have something to say about this becoming a long-term venture.

Stepping out of the story for a moment and before we move on, evidence shows that audio/video with a whiteboard is one of the most effective ways to communicate. The following graph compares the effectiveness and clarity of different models of communication.

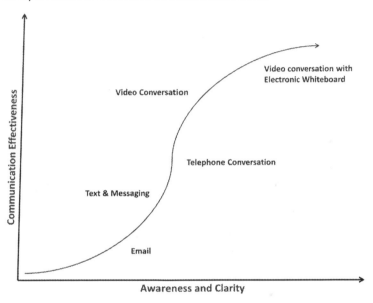

When Bill got a little arrogant, the screen took on a noticeably orange hue. As his mood changed and lightened up, so did the screen. This example with emotions of aggression as red and discernment as yellow had a blend of the two emotions—aggressive discernment. Therefore, the screen appeared orange.

They elect to schedule a follow-up where everyone can participate. The dates and times that all the participants are available appear as they agree on a time to continue. Each person has his or her own calendar, selecting warning times before terminating the interaction. Now, let's go back to the story.

Bill adds Mark to the internal portion of the strategy meeting with Sam. His keyboard disappears, and toward the bottom left, Bill's calendar is indicating an event coming up, so he then taps on his calendar. Just from tapping on the calendar, a previous interaction appears and indicates that Bill has a meeting with Tom Jones a few blocks away. Bill sets his iPad down at his desk in the holder and heads to the elevator to go to the garage. While he waits for the elevator, he opens up his new mobile device.

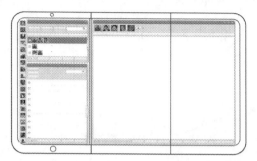

The company just gave Bill his mobile device this morning. It is a new tri-fold device that is just like a tablet when opened up. It is the latest technology that allows his screens to fold—a mobile unit that fits in his pocket can be opened into a tablet.

His last interaction that just completed with Sam is still up, and Mark is now present and utilizing audio/video. He asks Mark Smith to research a development for the conversation/conference that just ended. Mark has an air about him that seems to have everyone around him always smiling. Bill can tell that Mark is very pleased to be asked to get involved. Bill thinks, *I wonder why I hesitated to ask Mark. I really like him.*

Just then the elevator door opens, and he steps in. Looking in the mirror, he sees that he has a smile on his face. Bill is alone in the elevator. He closes his mobile device. In the elevator, a screen near the door of the elevator displays a map to his car, and he is asked, "Do you want to make a stop before going to your car?" Bill touches the No. "Do you want your car to be warmed up for you?" Bill is already used to each screen he comes to, relating to him in a way consistent with his current needs, immediate actions, and appropriate interest. He really likes it that each screen is aware and connects with him in the moment that he brings his attention to it. The door opens, and Bill exits.

His mobile device is closed into a single, small screen.

Bill looks at it as he walks to the car. Mark agrees to do the research and asks for a clarification. Bill responds as he walks to his car. Another thing Bill really likes about this new technology is that what Mark is scheduled to accomplish for this developing project is now in the realm of Bill's awareness; at the first indication of a change in Mark's plans, Bill will be immediately aware of the impact on this potential project.

When he enters his car, Bill puts the mobile device in his pocket, and his car immediately provides him with a GPS screen and an interaction. Bill thinks, *I can't get used to this. This technology is very intuitive. How in the world can my display in the car not be of my interaction with Mark? This new technology is really anticipating my needs. The current status of the interaction of the meeting I am going to is what is present.* It is an ongoing interaction for an entirely different interest from the one in the office with Mark. It is the meeting Bill is headed to right now. The GPS already has the best route to his meeting laid out.

Bill's company gave all vice presidents and higher new cars that no longer have dashboards or screens below the visual plane. With this new technology they concluded that instrumentation below the steering wheel takes excessive attention away from driving.

This new Corning technology allows for a semitransparent display in the windshield. Bill is pleased that he didn't have to settle for the alternative, a device built into the windshield blinder. Bill doesn't have to be concerned about that. He has his screen built into his windshield.

As Bill reviews what will take place at the meeting, his wife, Alice, appears and wants to go out for dinner tonight. She seems even more upbeat than usual. She wants to know if he has any preferences and what time he will be home. In his periphery, he notices that Alice's suggestion brought up several restaurants near their home. She says, "Italian." Three restaurants appear, and she selects Mario's. Bill suggests meeting at the restaurant at 6:00 p.m. A verbal response suggests a fifteen-minute warning, and Bill says, "Twenty-five minutes also." Once Bill and Alice conclude their conversation, the menu and map shrink, as if sucked into the calendar.

Meanwhile, where the conversation with his wife had shared a space with the pending meeting, this interaction now fills that space fully. Bill now considers some ideas, preceding his meeting with Tom Jones. When he pulls into the garage, the map shifts, and a parking place is identified. He is notified that he is early for his meeting. As he parks, the parking map disappears, and the path to the elevator is visible.

Before Bill enters the elevator, he is notified, "Tom is on the fifth floor. There is a Starbucks in the lobby. Do you want to get a drink before heading up?" Bill answers affirmatively and enters the empty elevator. A screen inside the elevator continues the interaction, and he contacts Tom to ask if Tom wants something brought up for him. Tom says he would like a Tall Pike with one teaspoon of sugar just as Bill connects with Starbucks. Bill places both of their orders with the Starbucks personnel, charges it, and then views the map on the elevator monitor showing the store location. The door opens, he turns the corner as the map indicated, and both drinks are ready.

Back to the elevator, he joins a passenger. An indicator with Bill's initials appears on the screen, prompting him to touch with his thumb if he wishes to interact. Bill maneuvers Tom's coffee and touches his left thumb to the display. On the display screen, Bill sees, "Tom is aware that you are heading up." Tom is turning the corner when the elevator door opens on the fifth floor. The two men greet each other, and Bill hands Tom his drink. Tom says, "Maybe I am old fashioned, but I appreciate that you have come to our office to finalize this." They head to the conference room, which has a whole roomful of people with screens in front of them. People both physically and virtually present are all settling in to continue an interaction that has been taking place over the last few weeks.

While Bill and others are meeting, the screens in front of them constantly give feedback on their current emotional states. Bill has designed the screens he interacts with to turn blue when he is fully connected, red when he is aggressive, and yellow when he is withdrawn—both in his thoughts (from visual clues) and in regard to what others are expressing. The screen darkens if he doesn't like something and brightens when he does. No matter where he is, screens are always depicting Bill's current emotional state. This awareness allows Bill to constantly shift and enhance his communication as he is conversing and interacting.

From his screen, Bill becomes aware of plans as they are discussed and how they might affect his company's resources, cash flow, and profitability. Analysis and speculative analysis are immediately available. This is true for all participants, giving parties from many different organizations a coordinated means to speculate about a shared project and find what best serves all organizations. Bill and others might be separately involved, or they might receive coaching and input from different departments and accountabilities. These meetings subsequently become extremely efficient, focusing on individual opportunities, as well as mutual benefits simultaneously.

Bill elects to get with his project manager, James. He likes that James is a no-nonsense manager. They communicate via text only as they engage in a private discussion as everyone speculates on negotiating the venture for a conclusion. One of the suggestions seems to impact Bill's company negatively regarding finances and available resources. Bill and James look at a few different options. They find one that actually improves profitability and reduces the demand on resources.

The agreement remains the same, increasing in value by shortening the time frame and the date of reimbursement. The commitments diminish slightly. He proposes the new guidelines, it works well for all the other parties, and the change is applied. Within twenty minutes, they are through, and Bill has timed his dinner with his wife perfectly.

A communication and accounting technology

Let's recap what just occurred. What you are viewing in this story is simply three paradigm shifts. The areas shifted are the following:

- ✓ Technology as a place to dwell as opposed to tools being used:

    - o   This new space is interest centric, not device or application centric

    - o   This technological space allows appropriate advertising to enter into your awareness (Starbucks, in this example)

    - o   People, resources, and support are available based on the interest and project

    - o   Where you are triggers an environment to support you:

        - GPS and maps—in the car, walking, outside, and inside (i.e., the elevator)

        - A continuous and fluid interplay of interactions, plans, events, and projects

        - Simple and informed choices based on all appropriate people/resources

        - Asks permission and authentication to interact when others are present in a public place

        - Appropriate ways of interacting—for instance, differing methods when a driver or a passenger is served

- ✓ Communication that is extremely effective

    - o   Self-awareness of your current emotional state helps maximize efficacy

        - Higher capacity to relate, connect, and communicate powerfully

        - Capacity to remain speculative much longer, increasing creativity

        - Support in asking for definite plans and commitments

o   Choices for methods of communication

- Text only

- Audio and text

- Audio/video with whiteboard

o   Consistent areas of interactions (interests/projects/events) that evolve over time

o   Maintaining only current/relevant conversations and decisions (which are nonlinear) that increase in simplicity and brevity over time

o   Ease of reviewing past and current changes (a historical review)

✓   Accounting—a new capability and design

o   Maintains coherence with current practices for tax and finance standards

o   Provides support and structure for agreements and commitments:

- They occur in the moment articulated, not as separate processes

- Negotiations and changes are fluid and immediate

- Future impact on resources, cash, and value is spontaneous and visible

- Collaboration is ongoing and organizationally designed with informed support

o   Provides a current future, present, and past in every moment

- Allows income (and expenses) to occur as interactions, and when completed, they become transactions of traditional financial accounting

- Allows three aspects for equity:

- Past periods via completed interactions or transactions

- Current status of completed interactions or transactions

- Future periods of equity, assets, and liabilities as they relate to the current state of interactions

- Shows current future value and present value each moment

  - Which is made available by accounting (of finances and resources intrinsically expressed) within each interaction

  - Each change, agreement, schedule, and plan is accounted for

  - Speculative ideas (and the potential influence on future finances and resources) can be examined easily

  - Strategies shift based on current events

- Three schedules of accounts provide the basics for an interactivity accounting

  - Income, which is Revenues – Expenses

  - Equity, which is Assets – Liabilities

  - Value, which is Agreements – Commitments

o Built on the premise that all transactions occur within interactions with

  - Persons;

  - Resources; and/or

  - Electronics

Can you spend a few moments with accounting, since it was only discussed briefly in the story?

Our current notions of accounting consist of the following:

- □ Where different points of completion and various transactions are recorded

- □ Where purchasing, selling, payments, receipts, and arrangement for payments and receipts exist

A future dimension is available in which arrangements for purchasing and selling gain a powerful footing.

Accounting becomes very simple. Management accounting and project, sales, and supply management used to be processes that required attending, coordinating, and posting. In contrast, are moments conversing. What used to occur as management system entries and planning projects one step at a time now takes place within interactions themselves.

Time, commitments, and agreements come into existence, change, shift, and go out of existence. All of this takes place instantaneously within a structure. Processes literally disappear. Because of this new technological, communicative environment, processes for training, approval, verification, and accounting no longer exist as the structure, and the arrangement of management accountability becomes simply a design function that occurs in an interaction.

We will soon enter into a world where technology provides an environment, much like different rooms of our home, where we live and work with our attention on what is important. Our interests are fully attended to and supported in this space. As you can see, this environment is not an application or tool. If applications or tools still exist, they will exist inside of this virtual environment only in ways that support what is presently important and our plans for the future. The world of linear thinking, e-mail, systems, applications, and devices that exist outside of our immediate interests will soon be history. We will welcome advertising that is specific to our actions and put forth at times when we are considering actions that forward both the advertiser and our own interests.

Implementation—a story

That was great! Can you give me another example— maybe someone who Bill's character is relating to—so I can gain more of a perspective?

Let's visit a portion of Bill's wife's life. Alice has obtained a very simple version of this new world. She got it to support her husband, who obtained his through his employer. Let's imagine that Alice's place of employment is incorporating this technology today, and she is supporting personnel in moving from e-mail to this virtual companion.

> Alice is finishing up with Jim, an Auto Zone sales representative and installer. He has just installed a rearview mirror and steering wheel device that gives Alice a new refit for her personal vehicle. She is the human resources director for a top talent agency, and her office is only ten minutes from both home and the kids' schools, very convenient for a working mom.
>
> "Okay, Alice, that is all installed for you; it's ready for you to try out. Go ahead and start your car," Jim instructs her. The rearview mirror not only gives her a view behind her, it also has many of the car's indicators built into it.
>
> "Jim, this mirror is a little bigger than my old mirror."
>
> Jim responds, "Yes, this has more height, but it won't obstruct your ability to see. If anything, you won't have to look down at your dash anymore, making it safer for you to drive. Where your left hand is—can you feel that? That's where some simple controls are. The top button activates the sound support, so you can talk to the mirror and give it instructions that might be out of the ordinary. The bottom button allows you to toggle between interactions if you need to. Are you ready to test it?"
>
> Alice has a twinkle in her eyes as she rejoins, "Sure!"
>
> Jim closes her car door and pulls out his mobile unit. A moment later, Jim's face appears on the very left of the mirror. Alice's mouth parts in astonishment as Jim's face in the mirror begins to move.

"Alice! Can you hear and see me okay?"

It takes a brief moment for Alice to regain her composure enough to respond, "Yes, this is amazing! Is there anything else that I need to know?"

Jim waves from outside the car window before he turns and heads toward the door, stepping back into the store. "Nope, it is just like your mobile device and the other small screens that you have already been using. There is nothing new to learn, and everything will be related to what you are doing. Janice from your office is supposed to be here in five minutes."

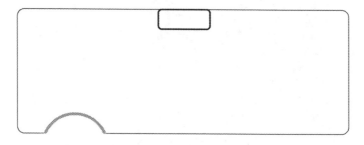

Alice has backed out of the parking place and is turning right onto the street when the mirror says in a very pleasant voice, "You are early heading back to the office. There is an office supply store on the way. Do you want to get that coloring book for Kate?"

Alice responds, "Yes, that would be great. Thank you for your help." In Alice's peripheral vision, the mirror shows two buttons and the top one is flashing. Alice presses the top button on the steering wheel once, and the flashing turns to a solid color. "That sounds good, yes. Connect me, please." A GPS appears, and the voice tells her to take the second right onto Redwood. She realizes that a man's face appears in the mirror.

"Hi, my name is Stuart. Welcome to Office Plus! How can I help you?"

"I need a coloring book for my daughter who goes to Stephen's Preschool. She says it is a Winnie the Pooh coloring book. Can you find that?"

Stuart responds, "Yes. It is $2.98 plus tax. You are a couple of minutes away. Do you want me to have it waiting for you at the front desk?"

She turns onto Redwood, and the GPS continues. In six blocks, she'll turn left onto Orleans. "Yes, Stuart. I'd appreciate that." The bottom button on the mirror flashes.

After Alice presses the bottom button on her steering wheel with her left hand, she hears Janice inquiring, "Hey, Alice, is Auto Zone on Brooks?"

"Yes, it is two blocks past Redwood on your left," Alice responds.

"Oh yes, there it is! Thanks."

Alice instructs, "See if you can get Marcus or Jim to help you. They're both really great. Be sure to tell them that you are brand new with this and get all of your questions answered." The bottom button on the mirror is blinking again. "It is really very intuitive. It is new, but there isn't really much to understand, Janice. Hey, I need to get this call. Let's talk at …" Alice glimpses at the screen on her mirror and notices that 3:15 p.m. and 3:30 p.m. have appeared. "Three thirty about Harrington. Okay?" There is a pause.

"Yeah, sure, that's perfect. Bye."

Alice presses the bottom button on her steering wheel, and Stuart says, "Oh, there you are. I have the coloring book ready for you."

The GPS says, "Office Plus in on your right."

Alice pulls into the parking place. "I am just pulling up, thank you," she says, finishing the call.

Getting out of the car and turning off the ignition, Alice opens the door and almost effortlessly floats up to the desk. From her mobile device, she gives Stuart access to her credit card account for the $3.23 charge. He smiles shyly. Stuart looks down briefly, picking up the coloring book and handing it to her while stuttering, "Th-thank you!"

She grins. "Thank you, Stuart." She gracefully turns away. Her grin gets wider as she opens the door and heads for her car.

A few minutes into Alice's next trip, the car mirror shifts the GPS to the right slightly and a ten-minute warning alerts her about a 2:00 p.m. meeting. The meeting is with her boss, John, regarding the support of Harrington's group on the transition project. Assured that nothing has changed and she is prepared, she pulls into the parking lot. She sees John isn't busy, presses the bottom button, and says, "I want to speak with John."

John says, "Hi, Alice."

"John, I'm a few minutes early, and I'm thirsty. Do you want anything from the soda shop downstairs?" Alice smiles briefly.

Smiling back, John says, "I'll have it ordered. Will you bring mine up with you? Do you want to get a bit of an early start? I'm ready for you."

Alice turns off the car, opens her mobile device, and says, "Sure, I will see you in a few minutes."

Upon opening up the door and turning immediately to the right, she sees a Sprite and a Coke sitting on the counter. The clerk returns Alice's smile from behind the counter and says, "John took care of these. Have a great day, Alice!"

Alice smiles in appreciation and says, "Thank you ... (glancing down to her mobile device) Fred. You too." Her mobile device is put away, and the drinks are taken up in each hand.

The elevator door opens, and the screen inside says, "Hello, Alice. John's office, is that correct?"

Alice grins again, still not used to this new screen in the elevator. "Yes, thank you."

The door opens, and John greets her with one of the biggest smiles she has ever seen on her boss's face. Reaching for his Coke, John says, "Alice, I don't know where to begin." He pauses, letting Alice join him before he turns to head to his office. "I don't think I have ever felt so much peacefulness, and Alice, really, I have rarely ever gone through a whole morning without having to reschedule at least a dozen things. Guess how many things I rescheduled this morning?"

Smiling, Alice says confidently, "None!"

He nods his head and looks at Alice seriously. "Do you know what that means?" He can't maintain his stern look, and a smirk appears. Alice raises her eyebrows as a reply. "It means you, once again, have delighted me beyond my expectations. I was really hesitant about this move away from e-mail and everything else." He gestures from behind his desk, inviting Alice to sit across from him and handing her a tablet. He adjusts his. Looking up, he continues, "Alice, there are no words that could possibly express my gratitude for you convincing me to not postpone this until the end of the month."

John adds Mark and Harrington to the interaction, even though both are not currently available, and suggests that they begin. Together, they discuss ideas regarding the transition. Alice brings Janice into the interaction, even though she knows that she is currently getting her new rearview mirror at Auto Zone. She and Janice will discuss things after she picks both kids up from school. The 3:30 p.m. get-together stands out on the calendar.

John and Alice each contribute to the planning on the whiteboard, scheduling this project out for the following week and planning for all the different support. Captured ideas are dragged off the interaction and into the calendar, keeping their work area clean and available for further planning. They finish fifteen minutes early, and both smile. "I have listened to so much more music today, just to fill in my free time. I'm a Sarah McLachlan fan; I know you like her too. I just discovered Dar Williams too. She's really great. Do you want to hear one of her songs?" John grins as he waits for a response.

Alice brightens. "Sure, I love your taste in music."

John touches the entertainment interest, then music,

then Dar Williams. "This is called 'The Beauty of the Rain.' I'd also like to play Natalie Merchant's 'Beloved Wife,' and then off with you."

Alice nods her approval as music filled the room.

Leaving work in a leisurely manner, Alice sits in front of the prekindergarten school a few minutes early and waits for Kate's school bell to ring. In the mood for some music, she picks a song to listen to. She reviews what she wants to go over with Janice in a half hour while relaxing to the music.

Kate struggles to open the back door, climbs in, and closes the door. She gets in her car seat, putting on her seat belt. Alice turns back and smiles, and Kate giggles.

"Hi, Mommy!"

After turning back around and putting the car in drive, she calls, "Hey, sweetie, can we keep listening to Mommy's music until we pick up Jake?"

Looking through the rear mirror, Kate says, "Sure, Mommy. Hey, did you get a new mirror, Mommy?"

Alice winks. "Yes, I did—a really fancy mirror that does a lot of things."

As they pull into the street, Kate says, "Cool!"

Jake has some comments about the mirror as well, and newly selected kid's music fills the car. Just before turning onto their street, Janice appears, and the button flashes. "Hey, guys, Mommy needs to take this." The music stops as Alice presses the button, and Janice begins speaking.

"Alice, this is so cool. I have never been so comfortable and felt so supported at work. It's like I never left home." Laughter filled Alice's car. "Hey, give me a minute. I'm home. I'll catch you inside my house." As the garage door closes behind them, the car door opens, and the kids run screaming into the playroom. Alice goes into the kitchen and picks up her tablet so she and Janice can continue.

After concluding with Janice, Alice touches "Family" on

her screen. A list with five headings appears on the left. She touches "Plans and Activities," and the calendar appears next to it, replacing the list. Alice sees that her husband, Bill, is available and in the car. "Oh, perfect! Hey, doll, can we go out to eat tonight? I want to celebrate! When will you be home?"

Bill says, "You sound even more excited than usual. Let's meet somewhere at six. What kind of food do you want?"

Alice thinks for just a second and says, "Italian!" Instantly, the calendar shortens by half and moves down as a list of three Italian restaurants near their home appears. "Mario's. Let's celebrate there. I'll see if Sandy or Debra can take care of the kids."

Bill says, "Either way, I will see you at 6:00 p.m. Can't wait!"

Seeing Bill's grin makes her face flush. She can feel the warmth in her cheeks. This fall, they will celebrate eleven years of marriage, and she has never been happier. Next, Alice drags Sandy and Debra onto the interaction space and types, "Can one of you come over today at 5:30 p.m. and stay with the kids until 8:00 p.m.?" Alice thinks of how great it will be when everyone is using this new technology. Then she will know if they are available or not. She looks at her choices ... text or e-mail. She touches text and sighs.

It has been a while since Alice came home wanting to unwind. She recalls the afternoon with John and the music they listened to, clicking on work and 2:15 p.m. Who was that singer? Ah, oh yes ... Dar Williams. She touches the name and then the interest "Entertainment" on the far right. She touches "Music" next, glancing on the right and selecting "Music Store" under resources. She picks iTunes from the selections and begins reading about this new singer/writer. She remembers Williams's *The Beauty of the Rain* album and buys it, while looking at the other titles. She thinks, *Oh, I really liked this one!* as she previews "The One Who Knows." She likes it too and buys it. Oh, there is the third song John played— "Farewell to the Old Me." Closing iTunes, "Music (Stuff)" now contains all the tunes that Alice has ever selected. Putting the tablet into its cradle, she walks into the media room, pulls out her mobile device, and touches the three songs she just bought. As she leans back in the lounge chair, the noise of the kids in the playroom falls away as the wonderful music fills the room.

Into the third song, "Family—Plans and Activities" opens with a text response from Debra: "Sorry, can't today."

Alice types back, "Thanks for getting back to me so quickly." She closes the screen, starting the music again. She hears her children playing in the background when the song ends and then chooses random songs from Sarah McLachlan and Natalie Merchant. Music begins playing once again. She turns on the big screen by enlarging the media room window and touching it to activate.

Her mobile unit and the big screen are working together. Touching the mobile device affects the big screen. Another text appears as the "Family" space opens up. It's Sandy.

"Sure, I can be there at 5:30 p.m."

Alice's mobile unit becomes a keyboard, and she types back, "Great! I will see you then. I will put some spaghetti and meat sauce out for you and the kids. Does that work for you?"

With the music filling the room, everything that gets added onto the big screen is displayed in a linear way.

"Sure, I love spaghetti and meat sauce. See you at 5:30 p.m."

Pausing the music, Alice gets up and walks around. She glances into the playroom, then moves on to the kitchen to get water and the tablet, bringing them back to the media room.

Alice smiles as one of her favorite songs, a Sarah McLachlan song that John introduced to her, began to play. She touches "Work" and then "Transition" and 3:30 p.m. on the calendar. Her entire conversation with Janice about Harrington comes up. As she looks at the whiteboard and the simple diagram, the screen grows dimmer and turns yellow. She knows what this means. She was becoming serious and didn't like what she was anticipating. Alice thinks, *How could this possibly happen while listening to Sarah McLaughlin's "Love Comes"?* While she grins at her humor, the screen brightens, and the background clears for just a moment before dimming and yellowing once more.

Alice sees the 2:00 p.m. event, "Transition Management," and, touching it, opens it also. The big screen and the tablet are identical, and if she touches and types on the tablet, the actions appear on both screens at the same time. The 3:30 p.m. and 2:00 p.m. interactions are sharing the right side of the screen. Alice highlights a portion of to 2:00 p.m. meeting that pertains to Harrington. She then opens up a window within the 3:30 p.m. discussion so that John can observe, and—if desired—she can have a private discussion with John before this interaction reconvenes. Inside the little window, she types, "I have some concerns about Harrington and Janice. I don't know exactly what is bothering me, but something is. Do you have any thoughts?" The screen clears and brightens even before Alice is done typing. *We will tackle this tomorrow,* she thinks. She closes up "Work" and relaxes to the music.

Later, while hanging out with the kids, the screen in the playroom notifies her that Sandy is due shortly. "Hey, kids, guess who is coming over? Sandy! Mommy is going out to dinner with Daddy; we'll be home before bedtime. Spaghetti night! Take your baths and everything before we get home, okay?" The doorbell rings, and the screens shows Sandy at the front door. "Hey, Sandy ... just a minute!" Alice dashes out of the room, heading for the front door.

Garage door opened, seat belt fastened, car slipped into reverse, she backs out of the garage just as she has a thousand times before. Driving through the alley fairly slowly, she has fifteen minutes to get to a restaurant less than ten minutes away. The bottom button blinks as her husband Bill appears, pushing the GPS over to the right. "Hey, sweetheart, see you shortly. I am halfway there." Alice smiles and says, "I will be there a few minutes early.

Do you want veal parmesan?"

"Yes, with some red wine," Bill says and then pauses. "You said we're celebrating?" Puzzled, as he notices that Alice has a frown on her face. The frown lifts as she replies, "Okay, that sounds great. Love you!"

Alice's smile could melt a glacier in the Arctic. She presses the upper button and says, "Get me the restaurant."

A pretty, oval-faced young girl appears. "Mario's … this is Maria. How can I help you?"

Alice begins, "My husband and I are on our way. Can I speak with our server?"

Maria responds with enthusiasm, "I see that Michael usually serves you—just a moment."

A familiar face appears. "Hello, Alice. I hear you and Bill are on your way. Does Bill want his usual?"

Alice smiles, "Yes, but we both want some red wine tonight."

Michael responds with, "Did someone get a raise?" He chuckles. "What do you want to try tonight? We have a special—a mixed seafood dish with a white wine garlic sauce."

"Is it really good?"

He smiles at her question. "It has a similar sauce to the one on the artichokes that you liked. I'd recommend it with ziti and a side of marinated asparagus."

"Oh, Michael, you're the best. That sounds wonderful! I will be there in a few minutes, and Bill will be there by 6:00 p.m." Michael's grin widens. "I will have your favorite table ready for you. See you soon." The GPS notifies Alice that she needs to turn onto Lemmon Ave in three blocks, as if she needs any help. She presses the right button to turn off the GPS.

This fabricated story stands in direct contrast with Alice's situation much earlier in the book. Usually, transitions are fraught with difficulties.

Though there will be enormous changes taking place for her and her workplace, in this transition from a world of processes to the structuring of relationships and dwelling in an environment specifically designed to support interactions, any difficulties that appear will occur within the support structure itself. Setup occurs in the interactions themselves. Just like moving into a new office or house, there is certainly a reorientation. However, inside an environment with the support of any resources that are needed appear the moment needed. In the earliest moments of this transition, the choices are about which resource might be the most supportive for each particular area being dealt with. Everything is resolved inside this technological space. This environment provides entry into everything that has existed and will occur in the future. It will be interesting, the choices made each moment, of adapting old furnishings (resources) or choosing new ones.

There is always continuity with this virtual companion. The restaurant, Mario's, once selected, was then the destination for both Alice and Bill. There was no need to enter it into the GPS. If a menu had been required, one would have been provided. When calling to the restaurant, Mario's hostess was instantaneously found. There were many other things not directly mentioned. How did Michael know Alice and Bill so well? How did he know that Bill ate the same thing while Alice was always trying something new?

The historical interactions between customers and suppliers or other such relationships give capacities beyond what memory generally provides for most of us. Would we become cynical in this new world, thinking that all terrific service was the product of living and working in this technological space? Sure, we could, but the screen would dim and turn yellow while we sulked. In this world, we simply won't be able to escape self-awareness (unless we actively choose to).

Development—a joint venture

I have a sense of how our activities and plans will be supported by the presence of this support system. Who is going to develop this?

That is a great question! It's also a very involved question. This invention has patents pending, which protects the intellectual property and the impact that this invention might have (see the last chapter, "Notes—Technical Mechanism Briefs" for more about patent pending). I anticipate that companies like Apple, Microsoft, Google, and others might be interested in licensing and developing applications that control the screens and desktops of devices. These types of companies are a logical choice because they are involved in areas like networking and advertising. I also anticipate that these enterprises and others are likely to step forward, as they have demonstrated social responsibility.

In the transition to interactivity accounting, spatial relativity technology, and the new ways we will be relating to ourselves and others, this invention might unavoidably disrupt various businesses and enterprises. There will likely be an enormous amount of new activity related to this new world we will dwell in. It might be very disruptive and controversial at first.

What do you mean by "disruptive and controversial"?

Historically, there have been many disruptions in the status quo. A few of these were the printing press, automobiles, electricity, and computers. Existing industries and vocations were displaced. When the typewriter disappeared as the computer arrived, companies like IBM had to shift to remain relevant in the world. With the automobile, the vital relationship with horses was disrupted, and early cars were not well thought of.

> Nowadays, the process of growth and development almost never seems to manage to create this subtle balance between the importance of the individual parts, and the coherence of the environment as a whole. One or the other always dominates.
> —Christopher Alexander[2]

This invention displaces several means of communicating, technological devices, and vast systems that support organizations in organizing and systemizing processes. A lot of employees serve functions like data entry, data retrieval, document preparation, and process management, occurring prior to and posterior to interactions. These activities will abruptly disappear. It is very likely that processes will be limited to interactions, affecting professional researchers and research and development departments in organizations. Outside of manufacturing and construction, most processes (busywork that currently takes place before and after interactions) lack any value in this new technological space.

Envisioning the future, I concern myself with these interruptions, but I also see the wonderful new industries and careers that could support interests and concerns for people, families, organizations, and governments much more effectively. I believe that this reorientation will be greatly enhanced by my invention. My agreement with licensing parties must embrace creative destruction and the development of very powerful structures for smooth transition as sizable portions of their investments.

The enterprises and those they rely on that develop this new structural technological space will likely be the first to benefit. Their reach and influence will undoubtedly powerfully ignite this technological evolution, even before it leaves its beta stage.

As our virtual companions quickly learn to anticipate our desires—our natural desire to be caring and our awareness of our capacity to be selfish or greedy—the enormous power that this invention will unleash will be counterbalanced. Having learned lessons from history, I am personally responsible for the new world we will be living in. Creating a network of support in the world we are stepping into is paramount. This world will seem unfamiliar, different from what we are used to, and it will invariably and unequivocally disrupt the status quo.

It is my intention that several well-established networks are in place and that the public is aware of these resources before the invention is introduced publically and in the workplace. Experts will support humanity as we transition. Those who participate as experts will likely be engaging in research. Speculatively, here a few domains of support:

- ✓ World religious and spiritual leaders—their interests, beliefs, platforms, and the best way to communicate

- ✓ Economic and political leaders—their theories, practices, privacy issues, beliefs, and the best way to communicate

- ✓ Powerful integrators, creative destruction consultants, transition experts, and the best way to communicate

- ✓ Organizations that creatively develop microprocessor devices, software (operating systems), networking, and the best way to communicate

Certain conditions exist, requiring research depicting the current condition of our world, regionally, in the domains of population, agriculture, industry, trade, health, wealth, infrastructure, culture, and politics to help determine what support might be needed most.

I think we can support a rapid entrance into a new world, based on my experience of humanity as emotional beings. I also trust that our nature is naturally loving and caring, and I have evidence supporting that conclusion.

What evidence do you have?

Several aspects of humanity support my conclusion that love is the basis for human life. One is simply how wild animals are more loving and peaceful once domesticated. Their repetitive interactions with humans alter their nature. Humans also seem to maintain a childlike quality throughout life.[3] The notion of beauty also influences humanity, as illustrated in our architecture, music, art, and poetry. As a student of discourses, I associate these broad themes that provide reality with the good and bad "forces" that are often referred to as morality and ethics. These are only a few examples supporting my proposition that human beings are loving beings.

I am concerned that this invention will especially influence those who enjoy competition, as they are often aggressive, influential, and financially powerful, but I also anticipate that organizations will be enticed by the simplicity and power that this new world will provide. The efficiency of organizations and the inherent impact on the bottom line couldn't be more pronounced. The pendulum may swing for a short period of time toward the benefit of the few or negative impact upon others, but this will be short lived. Regardless, this new world immediately opens up opportunities for all of us. As we gain the support that is readily available, we will derive financial benefit and vibrant opportunities. Within a few years, this new world could be available globally and begin to level the playing field. My guess is that by 2018, we will be living in a new world where humanity discovers not only the means but also the desire for all people to contribute and be contributed to, leaving no one aside.

If this new world is going to be so disruptive, why do you think we will sort things out by 2018?

One reason is the vigilance of this new model of communication and how it could influence all. When we reconstruct the paradigm of communication and dwell in this awareness, it will serve as a compass. We will become self-aware, seeing ourselves more clearly than ever before. We will be alerted when we are beginning to worry, being threatened, and the like. This compass will alter how we relate to ourselves. Much of what we seek will be immediately available to us. Possibly the most profound benefit is dissolving a delusion, giving humanity direct access to the source of our emotions and to what we experience as love, peacefulness, and happiness.

Our awareness reveals how our inherited notions of communication contribute to strife, war, genocide, and power seeking. This new model is and may remain controversial, even after the invention gives us a new world to live in. Most shifts of even smaller magnitude were ill received initially. Adopting this new communication model, however, isn't a prerequisite to the effective living afforded by my invention.

Having said all of that, I hope you connect with my experience that we are loving beings at some level because the power unleashed by this invention is enormous. Being immersed in this environment might, in part, be contrary to what feels normal, at least for a short period of time. I promise to be accountable for creating support for the possible creative destruction as well as the benefits that this invention might bring. This is my personal responsibility.

Equilibrium and other effects

Some people might not be as interested in benevolence as others. Do you think we will reach equilibrium?

Judging by our actions, it seems that our main desire is to experience love, peace, and happiness. All of our interests and concerns in life indicate this in one way or another. I like that you are using the term equilibrium. The concept of equilibrium is at the heart of this new technological space and the new paradigm of communication.

From a communication perspective, it appears that three levels are occurring in unison: love, aggression, and discernment. Love is a relational space with stillness and peace, timelessness, the joyful experience of wonder, and light. Aggression is an experiential, "now" phenomenon, active and expressive. Discernment allows for continuity, movement, understanding, and awareness. If we view one of these levels as more valuable than another, we miss the point of balance. Without discernment, for example, appreciation of experiencing love wouldn't exist!

Yes, reaching equilibrium, with our hearts as our guide!

In reflecting on the notion of equilibrium, I recognize that I must be responsible for my bias and predisposition for my association with being related, the expressions of my heart. Others might discover a different nuance that acts as a guiding compass. The onus on to each of us is to discover what inspiration brings us powerfully into the world.

All three levels seem to be available each moment, and the opportunity is for a rich balance of all of them. We may find that one influences us more than another, given our uniqueness. The three levels are of the heart (relating), of experience (the poignancy of "now"), and of word (integrity).

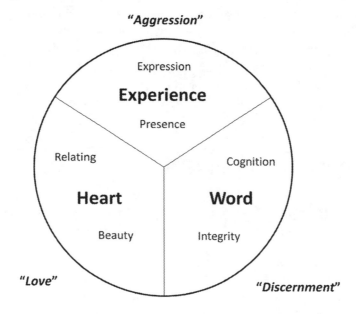

Conclusion

One of the first questions you asked me was, "Is this book just about introducing an invention?" I would like to address that question a bit more now. The purpose of this book is to illustrate and give notice of an invention that will provide an explosive expansion of our capacities. The invention was inspired by my desire for a completely new world in which we effectively relate to both others and ourselves. This new world will expand and empower all of our worldly commitments. I am counting on all of us to bring our love for life forward because what we devote our lives to will come with power and ease.

We must ascend toward personal responsibility. Only then will our dreams and aspirations be acted on. Simplicity is the key. I attempt to organize my life in such a way that all important areas are fully taken care of. Why? Because I am committed to experiencing life as love, peace, and joy. When I focus on the important areas, I spend less time on what I am not committed to. I make plans and take actions that support my family, friends, work, career, education, money, play, wealth, health, sexuality, spirituality, and other interests.

Everyone has aspirations, colored by each of life's unique experiences. In the absence of all turmoil, love, peace, and joy abound. What provides this abundance doesn't disappear. Rather, it resides eternally underneath the surface of experiences, actions, expressions, and thoughts. Each of us has a gift to express, and part of life is discovering what our life's purpose might be: what we have to contribute to humanity. This invention brings clarity and gives us direct access to what is important and what contributes and helps us by giving us a compass for living a life we love.

You have been talking about the invention. What about this book?

I am glad you are asking this question. We must wait for the new technological space. However, you have spent the last few hours reading and thinking about communication and our existence. I think awareness of three distinct levels of existence provides a life-altering fullness.

It might be subtle. The next time you get upset with someone close to you, in the midst of an emotional experience of the moment, a background of connectedness might be present also. You might notice that each moment is more poignant and alive, or when it isn't, you will be aware that you are deep in thought and removed from life.

I also think that you will begin questioning all the theories and conclusions of philosophy and science, answering the riddles of life. Life is in the living, not merely understanding. It has been my experience that gaining awareness of understanding as the tiniest percentage of life's experiences is enlightening. Awareness is the ultimate gateway to freedom.

You might be in touch with clarity right now. If you are, consider that you did some valuable work. I wrote what I wrote, but you provided a rigor for thinking and attention to your experiences of life. You accomplished this. I conclude with a quote from one of the most talented and prolific writers in history: "It is with books, as it is [with fire]; one goes and lights his candle at his neighbor's, and then lights one of his own; [communicating with others], so that it becomes absolutely the property of every one."[4] In other words, if you found value in this book, share it with others.

1. Christopher Alexander, *The Nature of Order: An Essay on the Art of Building and the Nature of the Universe, Book 4—The Luminous Ground* (Berkley: Center for Environmental Structure, 2003).

2. Christopher Alexander, Murray Silverstein, Shlomo Angel, Sara Ishikawa, and Denny Abrams, *The Oregon Experiment* (New York: Oxford University Press, 1975), 12.

3. "Biology of Love," Humberto Romesin Maturana and Gerda Verden-Zoller, (Focus Heilpadagogik, Ernst Reinhardt, Munchen/Basel 1996), accessed May 22, 2012, http://www.lifesnaturalsolutions.com.au/documents/biology-of-love.pdf

4. Voltaire, *The Works of Voltaire. A Contemporary Version. A Critique and Biography by John Morley, notes by Tobias Smollett*, trans. William F. Fleming (Indianapolis: Liberty Fund, 2010), http://files.libertyfund.org/output/epub/Voltaire_0060-19p2.epub.

CHAPTER 11
NOTES: TECHNICAL
MECHANISM BRIEFS*

Network essentials

***Before the story, you showed a network diagram with a
brief explanation. Will you elaborate on this?***

A virtual (computer-generated) environment takes on an unambiguous
and explicit orientation. That orientation occurs as physical and corporeal.
A virtual world mirrors the physical world from the perspective of what
is important. Again, it is like a virtual reality mapping directly onto our
physical reality and offering support, such as people and resources, that
can help us.

This technological space occurs as a servant, a companion that shifts
simultaneously with our interests and circumstances. This companion
provides us with all the people and resources necessary to support what
is important to us. Over time, its capability to anticipate our desires grows,
and it encourages us to communicate in an effective manner, providing
feedback to assist us.

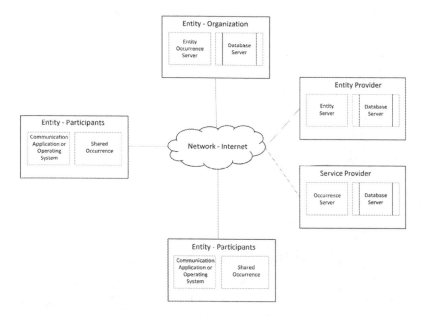

The image above depicts the most basic mechanism and functionality of this technological space. The entity participant, or you and I, rely on network-capable electronic devices that have at least screens, audio capacities, and other means for input or output. These devices require an operating system and/or software that hijack the functions of the device as to provide a space for the interactions.

This space gives access to the interests, resources, stuff, and interactions for the technological space. The devices simply coordinate the current space as it maps onto your current intentions. This space provides an occurrence that might be shared among one or more other participants.

There are also two or more types of providers. One is an entity provider, which is an entity server supported by a data server that learns about the entity: the wants, preferences, expressions, networks of support—the people, resources, interests, concerns, affiliations, and structures. Another is a service provider: an occurrence server supported by a data server. This provides a shared environment for interactions, media, collaboration, and agreements.

Will you describe the entity server more fully?

An entity provider is part of a network (cloud) environment, and its function is to manage, define, and continually maintain the aspects, relationships, preferences, distinctions, and propensities of an identity, such as structures, definitions, schedules, and templates. Additionally, all the stuff that is part of an entity's support and design structures are kept consistent by this capacity.

The entity provider works seamlessly with service providers. This coordination is dependent on the communications operating system and the interim software application.

The entity server provides relationship and orientation for the entity. It is specific to the entity, and the relationship inherited over time provides a coherent and consistent foundation for the entity. This is thusly a learning environment.

The entity server's ability to provide relevancy within an interactivity gives a correlation between the entity server and the occurrence server. It categorizes and holds a position for each interaction relative to the interests, projects, and relationships.

The purpose of the entity server is to provide controls and influence the look, feel, and the specific learning and specificity for each person or organization. Each entity identity is unique. Unlike a namespace, there is no association other than itself. The entity server connects each entity to others by interests and relationships with other resources which are many times entities with a specific identity. I might refer to these unique entities as entity domains.

What is an entity domain?

Entity domains are a technological physicality. They are a combination of microprocessor hardware, software, and available storage capabilities. Specific entities, identities, and contexts are specified in a unique way, not in the hierarchical way of current Domain Name Systems (DNS) but rather in a relational methodology. Another way of saying this is that entities correlate, connect, and coordinate via relationships, and those relationships are unique to each identity.

To illustrate this, entity 1 → relationship 1 → entity 2 is distinct from entity 2 → relationship 2 → entity 1. In action, entity 1 and entity 2 are unique and coherent, whereas relationship 1 and relationship 2 are dissimilar because they uniquely correlate with entity 1 and entity 2. To further clarify this, let's say that entity 1 is software engineer Pete Smith and entity 2 is software supervisor Tom Jenkins. Relationship 1 might have a hierarchical structure like work/company A/operations/team/boss and relationship 2 might have a hierarchical structure like work/company A/operations/team A/design A/minimal supervision.

You can begin to see the design implications here. It is the relationships or memberships with organized entities that provide a sense of meaningful structure, order, and coherence within specific interests.

What is an occurrence server?

An occurrence server contains elements that are part of an event that can be shared by two or more persons. The server provides the common aspects of the interactivity space. It coordinates effectively with the entity server, making private interactions and sensitive portions of public interactions unavailable to people who are not included. It specifies the participants in such a way that each interactive window is appropriate and can be limited based on participants.

The common occurrence is served up such that it is pertinent for each participant. Private interactions appear differently for different participants. Different titles might be present, and templates are matched to each participant. A template could provide a sales arrangement for one participant and a purchasing capability for another. It is the entity server that influences these differences.

The service provider (occurrence server) and the entity provider (entity server) coordinate to produce the displays and to allow the opening of activities by operating system for any device that is relevant to the entities involved in a shared occurrence.

Organizations have a hybrid of these two servers that permits a capacity to provide strictly internal identities like groups of employees, plumbers, and dispatchers as examples. It also gives local access to support systems over local area networks.

Why don't you ask a question that might tie this up in a simple example? Go ahead, ask your question.

Interfaces

You mentioned templates and systems providing translations for applications. Can you give a quick example?

Okay, let's say you are cowriting a book. You like Microsoft Office, and she likes iWorks. As long as you both have a template (application) and a system translator (if required), you both can simultaneously write the same book, in real time and/or at different times. The manuscript occurs on an occurrence service provider. You can add your editors, artists, and other collaborators. Each of you would view the same book from many different points of view (applications), depending on the template/systems chosen. It would occur as a project from different views also. The entity provider gives each person his or her specific preferred treatment and view.

The amount of time to write a book in this environment and have it published will be much more efficient. Your capacity for collaboration is only limited by your imagination. That is a simple example. Thank you for this interaction. I think it made a difference for those reading.

Please, one more question. If you were going to explain this invention simply, what would you say?

This is the last question (I am smiling as I write this). You really want me to tie this up in a little bow, don't you? I will attempt to provide an improvisation using analogies. One likeness might look like an elaborate e-mail server; however, instead of a database of linear e-mails, a temporal sequence or slices of altering interactions, common electronic whiteboard interactive formations capable of every means, such as audio/visual, text, videos, pictures, drawings, diagrams, and templates to interact. This analogy would allow for a sophisticated namespace that is intelligent, learning, and gains new capacities over time, while cataloging and organizing the communication space with a whole array of capabilities.

Another analogy would be from your viewpoint, where you are accessing an enhanced browser, except that the controls and extensive capacities all exist on a specific network controller, and the sites aren't really sites—they are different time slices of many communication happenings, controlled and being given support and tools by the controlling server.

These analogies both fall very short, but maybe they help in a small way. One other example might be a glorified Google Wave[1] that relied on software and had expanded features way beyond its design. It also didn't use blips or occur linearly or require an entry into an earlier blip. This environment would remain completely fluid instead without the boxy structure of blips (similar to other linear environments like Google+ and Facebook). Once again, this isn't a very good analogy. Social networks don't begin to capture this world either. Maybe it is safe to say, "It took a book to reveal this new technological world."

Thank you for your attention and interest. Until we experience it, we won't have a decent grasp of how this new technological world will alter our work and our lives.

This concludes the notes. Visit the website: http://dwellinginanewworld. com.

1. "Google Wave," Google, accessed May 22, 2012, http://support. google.com/wave/bin/topic.py?hl=en&topic=25155&parent=TopL evel&ctx=topic.

Afterword: Regarding licensing

To expand on the question in the last chapter—"What do you mean, disruptive and controversial?"—I am extremely certain of the creative destruction due to interactivity accounting. I really can't imagine someone attempting to design a complex logic to integrate interactivity accounting in a process-oriented, linear environment to circumvent licensing. The convergence of a novel technological space and the elimination of linear communication, however, empower a very simple methodology to come into existence.

It is my intention that the creative destruction of industries and employment be mitigated through contractual agreements. These agreements provide the methodology described in this book and within the patents pending. The enterprises that gain access to licensing will be asked, as an essential part of their agreement, to commit to funding a self-governing organization that will organize powerful and effective networks and services that include but are not limited to the following:

- ✓ Integrators, creative destruction experts, and transition consultants/services for developed societies

- ✓ Enterprises that develop microprocessor devices, software, and network-related technology

- ✓ A network for collaboration, developing standards that permit inclusion, coordination, and transparency (legal implications also)

- ✓ Engaging in research and establishing outreach for undeveloped societies in the following areas:

- o Conditions of population, health, and hunger (agriculture)

- o Industry, trade, and infrastructure

- o Education and finance

✓ Global involvement with peace initiatives and encouraging respect for differences

✓ Promotion of opportunities for creative use of military applications in peaceful practices

✓ Creation of opportunities to dissuade crime

The organization that will develop and manage this fund and projects shall be autonomous from the licensees. While they might have some representation, the licensees would not have a controlling influence. In addition, this organization will monitor and coordinate with

✓ world religious and spiritual; and

✓ economic and political leaders.